Nonplussed!

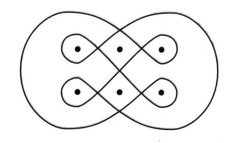

Nonplussed!

MATHEMATICAL PROOF OF IMPLAUSIBLE IDEAS

Julian Havil

PRINCETON UNIVERSITY PRESS

PRINCETON AND OXFORD

Published by Princeton University Press,
41 William Street, Princeton, New Jersey 08540

In the United Kingdom: Princeton University Press,
3 Market Place, Woodstock, Oxfordshire OX20 1SY

Library of Congress Cataloguing-in-Publication Data

Havil, Julian, 1952–
Nonplussed! : mathematical proof of implausible ideas / Julian Havil.
p. cm.
Includes index.
ISBN-13: 978-0-691-12056-0 (acid-free paper)
ISBN-10: 0-691-12056-0 (acid-free paper)
1. Mathematics–Miscellanea. 2. Mathematical recreations.
3. Paradox–Mathematics. I. Title.
QA99.H38 2006
510–dc22   2006009994

British Library Cataloguing-in-Publication Data
A catalogue record for this book is available from the British Library

This book has been composed in Lucida
Typeset by T&T Productions Ltd, London
Printed on acid-free paper ∞
press.princeton.edu
Printed in the United States of America

3 5 7 9 10 8 6 4

To Anne
for whom my love is monotone increasing
and unbounded above

Time flies like an arrow. Fruit flies like a banana.

Groucho Marx

Do I contradict myself? Very well then I contradict myself. I am large, I contain multitudes.

Walt Whitman

Mathematics is not a careful march down a well-cleared highway, but a journey into a strange wilderness, where the explorers often get lost. Rigour should be a signal to the historian that the maps have been made, and the real explorers have gone elsewhere.

W. S. Anglin

Contents

Preface

Epistle to the Reader

I HAVE put into thy hands what has been the diversion of some of my idle and heavy hours. If it has the good luck to prove so of any of thine, and thou hast but half so much pleasure in reading as I had in writing it, thou wilt as little think thy money, as I do my pains, ill bestowed. Mistake not this for a commendation of my work; nor conclude, because I was pleased with the doing of it, that therefore I am fondly taken with it now it is done. He that hawks at larks and sparrows has no less sport, though a much less considerable quarry, than he that flies at nobler game: and he is little acquainted with the subject of this treatise – the UNDERSTANDING – who does not know that, as it is the most elevated faculty of the soul, so it is employed with a greater and more constant delight than any of the other. Its searches after truth are a sort of hawking and hunting, wherein the very pursuit makes a great part of the pleasure. Every step the mind takes in its progress towards Knowledge makes some discovery, which is not only new, but the best too, for the time at least.

These words, recorded as being written in Dorset Court, London, on 24 May 1689, are those of the British philosopher and polymath John Locke and form the first part of his Preface (or Epistle to the Reader) of his monumental work of 1690, *An Essay Concerning Human Understanding.*

It is our preface too.

Acknowledgements

I should like to thank my headmaster, Dr Ralph Townsend, for his support, particularly through sabbatical leave, former student Tom Pocock for his enthusiasm and honest opinions, the reviewers for their helpful views, Design Science for creating Mathtype^{TM} and Wolfram Research for creating Mathematica^{TM}. Further, my grateful thanks are due to Jonathan Wainwright of T&T Productions Ltd for his meticulous and patient work and to my editor, Vickie Kearn, for her own patient understanding and enthusiasm. Finally, I join a long list of those who have thanked Martin Gardner for being a lifelong inspiration.

Nonplussed!

Introduction

⊙

Alice laughed: 'There's no use trying,' she said; 'one can't believe impossible things.'

'I daresay you haven't had much practice,' said the Queen. 'When I was younger, I always did it for half an hour a day. Why, sometimes I've believed as many as six impossible things before breakfast.'

'Where shall I begin,' she asked.

'Begin at the beginning,' the king said, 'and stop when you get to an end.'

Lewis Carroll

It does not take a student of mathematics long to discover results which are surprising or clever or both and for which the explanations themselves might enjoy those same virtues. In the author's case it is probable that in the long past the 'coin rolling around a coin' puzzle provided Carroll's beginning and a welcome, if temporary, release from the dry challenges of elementary algebra:

> Two identical coins of equal radius are placed side by side, with one of them fixed. Starting head up and without slipping, rotate one about the other until it is on the other side of the fixed coin, as shown in figure 1.
>
> Is the rotated coin now head up or head down?

Within a random group of people both answers are likely to be proffered as being 'obviously true', yet one of them is false and a quiet experiment with two coins quickly reveals which. We must prove the fact though, and too much knowledge is dangerous here: fix on a point on the circumference of the moving circle and we have an epicycloid to consider (or, more precisely, a cardioid) – and there could be hard mathematics to deal with.

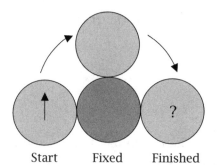

Start   Fixed   Finished

Figure 1. A coin rolling around another fixed coin.

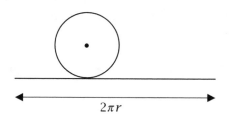

$2\pi r$

Figure 2. The situation simplified.

Alternatively, concentrate on the path of the centre of the moving coin and let us suppose that the common radii of the coins are r. During the motion, the path traced out by this centre is a semicircle, whose centre is itself the centre of the fixed coin and whose radius is $2r$; the motion will cause the centre to move a distance $\pi(2r) = 2\pi r$.

Now simplify matters and consider the moving coin rotating without slipping along a straight line of length $2\pi r$, the distance moved by its centre, as shown in figure 2. It is perfectly clear that it will have turned through 360° – and so be the right way up.

When it is first seen, the result is indeed surprising – and the solution clever.

It is a suitable preliminary example as this book chronicles a miscellany of the surprising, with a nod towards the clever, at least in the judgement of its author. The choice of what to include or, more painfully, what to exclude has been justly difficult to make and a balance has been found which recognizes the diversity of the surprising as well as the large role played by probability and statistics in bringing about surprise: it is they and the infinite which abound in the counterintuitive; other areas of mathematics dally with it. To reflect all of this, the fourteen chapters which constitute the book are divided evenly and alternate between results which intrinsically depend on probability and statistics and those which arise in other, widely diverse, areas; one such is the infinite. To reflect these tensions further, this is the first of two such books, the second providing the opportunity to embrace what the reader may have considered as unfortunate omissions. Wherever it has been possible, the provenance of the result in question has been discussed, with a considerable emphasis placed on historical context; no mathematics grows like Topsy, someone at some time has developed it.

Apart from chapter 13 (and where else could that material be placed?), the level of mathematics increases as the book progresses, but none of it is beyond a committed senior high school student: looking hard is not at all the same as being hard. It is hoped that the reader, young or not-so-young, will find something in the pages that follow to inform or remind him or her of the frailty of the intuition we routinely employ to guide us through our everyday lives, but which is so easily confounded – only to be replaced by the uncompromising reason of mathematical argument.

Chapter 1

THREE TENNIS PARADOXES

So that as tennis is a game of no use in itself, but of great use in respect it maketh a quick eye and a body ready to put itself into all postures; so in the mathematics, that use which is collateral and intervenient is no less worthy than that which is principal and intended.

Roger Bacon

In this first chapter we will look at three examples of sport-related counterintuitive phenomena: the first two couched in terms of tennis, the third intrinsically connected with it.

Winning a Tournament

The late Leo Moser posed this first problem during his long association with the University of Alberta. Suppose that there are three members of a club who decide to embark on a private tournament: a new member M, his friend F (who is a better player) and the club's top player T.

M is encouraged by F and by the offer of a prize if M wins at least two games in a row, played alternately against himself and T.

It would seem sensible for M to choose to play more against his friend F than the top player T, but if we look at the probabilities

Table 1.1. The situation if the new member plays his friend twice.

F	T	F	Probability
W	W	W	ftf
W	W	L	$ft(1-f)$
L	W	W	$(1-f)tf$

Table 1.2. The situation if the new member plays the club's top player twice.

T	F	T	Probability
W	W	W	tft
W	W	L	$tf(1-t)$
L	W	W	$(1-t)ft$

associated with the two alternative sequences of play, FTF and TFT, matters take on a very different look. Suppose that we write f as the probability of M beating F and t as the probability of M beating T (and assume independence).

If M does choose to play F twice, we have table 1.1, which lists the chances of winning the prize.

This gives a total probability of winning the prize of

$$P_F = ftf + ft(1-f) + (1-f)tf$$
$$= ft(2-f).$$

Now suppose that M chooses the seemingly worse alternative of playing T twice, then table 1.2 gives the corresponding probabilities, and the total probability of winning the prize becomes

$$P_T = tft + tf(1-t) + (1-t)ft$$
$$= ft(2-t).$$

Since the top player is a better player than the friend, $t < f$ and so $2 - t > 2 - f$, which makes $ft(2-t) > ft(2-f)$ and $P_T > P_F$. Therefore, playing the top player twice is, in fact, the better option.

Table 1.3. Outcome of an all-plays-all
tournament between the various teams.

$$T_B$$

		10	2	3	7	5
	1	B	B	B	B	B
	8	B	W	W	W	W
T_W	9	B	W	W	W	W
	6	B	W	W	B	W
	4	B	W	W	B	B

Logical calm is restored if we look at the expected number of wins. With FTF it is

$$
\begin{aligned}
E_F = {} & 0 \times (1 - f)(1 - t)(1 - f) \\
& + 1 \times \{f(1 - t)(1 - f) + (1 - f)t(1 - f) + (1 - f)(1 - t)f\} \\
& + 2 \times \{ft(1 - f) + f(1 - t)f + (1 - f)tf\} + 3 \times ftf \\
= {} & 2f + t
\end{aligned}
$$

and a similar calculation for TFT yields $E_T = 2t + f$.

Since $f > t$, $2f - f > 2t - t$ and so $2f + t > 2t + f$, which means that $E_F > E_T$ – and that we *would* expect!

Forming a Team

Now let us address a hidden pitfall in team selection.

A selection of 10 tennis players is made, ranked 1 (the worst player, W) to 10 (the best player, B). Suppose now that W challenges B to a competition of all-plays-all in which he can chose the two best remaining players and B, to make it fair, must choose the two worst remaining players.

The challenge accepted, W's team is $T_W = \{1, 8, 9\}$ and B's team is $T_B = \{10, 2, 3\}$. Table 1.3 shows the (presumed) inevitable outcome of the tournament; at this stage we are interested only in the upper left corner. We can see that W's disadvantage has not been overcome since T_B beats T_W 5 games to 4.

Table 1.4. The average rankings of each of the three pairs of teams.

Average ranking of T_B	5	$5\frac{1}{2}$	$5\frac{2}{5}$
Average ranking of T_W	6	6	$5\frac{3}{5}$

The remaining players are $\{4, 5, 6, 7\}$ and W reissues the challenge, telling B that he can add to his team one of the remaining players and then he would do the same from the remainder; of course, both B and W choose the best remaining players, who are ranked 7 and 6 respectively. The teams are now $T_W = \{1, 8, 9, 6\}$ and $T_B = \{10, 2, 3, 7\}$ and the extended table 1.3 now shows that, in spite of B adding the better player to his team, the result is worse for him, with an 8–8 tie.

Finally, the challenge is reissued under the same conditions and the teams finally become $T_W = \{1, 8, 9, 6, 4\}$ and $T_B = \{10, 2, 3, 7, 5\}$ and this time the full table 1.3 shows that T_W now beats T_B 13–12.

A losing team has become a winning team by adding in worse players than the opposition.

Table 1.4 shows, in each of the three cases, the average ranking of the two teams. We can see that in each case the T_B team has an average ranking less than that of the T_W team and that the average ranking is increasing for T_B and decreasing (or staying steady) for T_W as new members join. This has resonances with the simple (but significant) paradox known as the *Will Rogers Phenomenon.*

Interstate migration brought about by the American Great Depression of the 1930s caused Will Rogers, the wisecracking, lariat-throwing people's philosopher, to remark that

> When the Okies left Oklahoma and moved to California, they raised the intellectual level in both states.

Rogers, an 'Okie' (native of Oklahoma), was making a quip, of course, but if we take the theoretical case that the migration was from the ranks of the least intelligent of Oklahoma, all of whom were more intelligent than the native Californians(!), then what he quipped would obviously be true. The result is more subtle, though. For example, if we consider the two sets $A = \{1, 2, 3, 4\}$ and $B = \{5, 6, 7, 8, 9\}$, supposedly ranked by intelligence level (1 low, 9 high), the average ranking of A is 2.5 and that of B is 7. However, if we move the 5 ranking from B to A we have that $A = \{1, 2, 3, 4, 5\}$ and $B = \{6, 7, 8, 9\}$ and the average ranking of A is now 3 and that of B is 7.5: both average intelligence levels have risen.

If we move from theoretical intelligence levels to real-world matters of the state of health of individuals, we approach the medical concept of *stage migration* and a realistic example of the Will Rogers phenomenon. In medical stage migration, improved detection of illness leads to the fast reclassification of people from those who are healthy to those who are unhealthy. When they are reclassified as not healthy, the average lifespan of those who remain classified as healthy increases, as does that of those who are classified as unhealthy some of whose health has been poor for longer. In short, the phenomenon could cause an imaginary improvement in survival rates between two different groups. Recent examples of this have been recorded (for example) in the detection of prostate cancer (I. M. Thompson, E. Canby-Hagino and M. Scott Lucia (2005), 'Stage migration and grade inflation in prostate cancer: Will Rogers meets Garrison Keillor', *Journal of the National Cancer Institute* 97:1236–37) and breast cancer (W. A. Woodward et al. (2003), 'Changes in the 2003 American Joint Committee on cancer staging for breast cancer dramatically affect stage-specific survival', *Journal of Clinical Oncology* 21:3244–48).

Winning on the Serve

Finally, we revert to lighter matters of tennis scoring and look at a situation in which an anomaly in the scoring system can, in theory, be exposed.

The scoring system in lawn tennis is arcane and based on the positions of the hands of a clock. For any particular game it is as follows.

> If a player wins his first point, the score is called 15 for that player; on winning his second point, the score is called

30 for that player; on winning his third point, the score is called 40 for that player, and the fourth point won by a player causes the player to win, unless both players have won three points, in which case the score is called deuce; and the next point won by a player is scored 'advantage' for that player. If the same player wins the next point, he wins the game; if the other player wins the next point the score is again called deuce. This continues until a player wins the two points immediately following the score at deuce, when that player wins.

The great tennis players of the past and present might be surprised to learn that, with this scoring system, a *high quality tennis player serving at 40-30 or 30-15 to an equal opponent has less chance of winning the game than at its start.*

We will quantify the players being evenly matched by assigning a fixed probability p of either of them winning a point as the server (and $q = 1 - p$ of losing it); for a high quality player, p will be close to 1. The notation $P(a, b)$ will be used to mean the probability of the server winning the game when he has a points and the receiver b points; we need to calculate $P(40, 30)$ and $P(30, 15)$ and compare each of these with $P(0, 0)$, which we will see will take some doing!

First, notice that the position at 'advantage' is the same as that at $(40, 30)$, which means that the situation at deuce, when divided into winning or losing the next point, is given by

$$P(40, 40) = pP(40, 30) + qP(30, 40),$$

also, using the same logic, we have

$$P(30, 40) = pP(40, 40) \quad \text{and} \quad P(40, 30) = p + qP(40, 40).$$

If we put these equations together, we get

$$P(40, 40) = p(p + qP(40, 40)) + q(pP(40, 40))$$

and so

$$P(40, 40) = \frac{p^2}{1 - 2pq}.$$

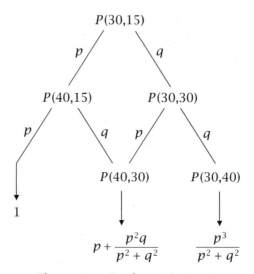

Figure 1.1. Finding $P(30, 15)$.

Using the identity $1 - 2pq = (p + q)^2 - 2pq = p^2 + q^2$ we have the more symmetric form for the situation at deuce,

$$P(40, 40) = \frac{p^2}{p^2 + q^2},$$

and this makes

$$P(30, 40) = pP(40, 40) = \frac{p^3}{p^2 + q^2}$$

and the first of the expressions in which we have interest is then

$$\boxed{P(40, 30) = p + \frac{p^2 q}{p^2 + q^2}}$$

Now we will find the expression for $P(30, 15)$, which takes a bit more work, made easier by the use of a tree diagram which divides up the possible routes to success and ends with known probabilities, as shown in figure 1.1.

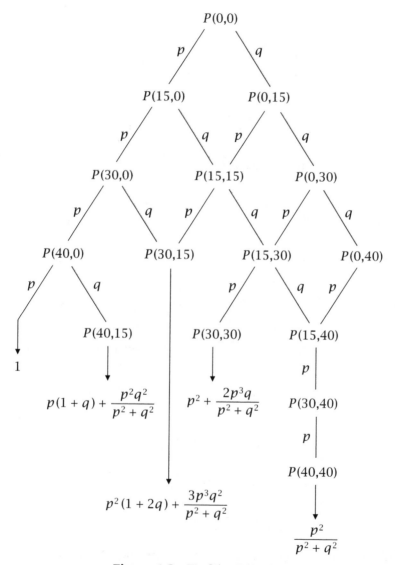

Figure 1.2. Finding $P(0,0)$.

Every descending route is counted to give

$$P(30,15) = p^2 + 2pq\left(p + \frac{p^2q}{p^2 + q^2}\right) + q^2\left(\frac{p^3}{p^2 + q^2}\right)$$

$$= p^2(1 + 2q) + \frac{3p^3q^2}{p^2 + q^2}$$

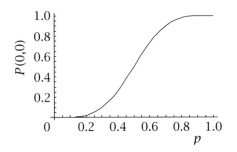

Figure 1.3. $P(0,0)$ plotted against p.

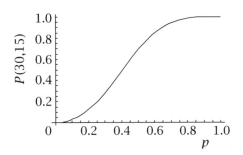

Figure 1.4. $P(30,15)$ plotted against p.

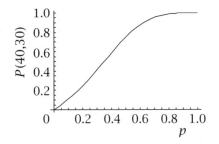

Figure 1.5. $P(40,30)$ plotted against p.

and so we have found the second of our expressions

$$P(30,15) = p^2(1+2q) + \frac{3p^3q^2}{p^2+q^2}$$

We only need the starting probability $P(0,0)$, which is by far the hardest goal, and to reach it without getting lost we will make use of the more complex tree diagram in figure 1.2, which again

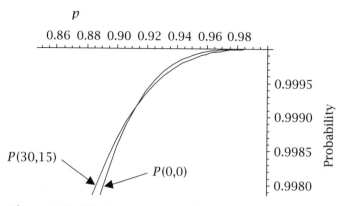

p

0.86 0.88 0.90 0.92 0.94 0.96 0.98

0.9995

0.9990

0.9985

0.9980

Probability

$P(30,15)$

$P(0,0)$

Figure 1.6. The intersection of $P(0,0)$ with $P(30,15)$.

shows the ways in which the situations divide until a known probability is reached. We then have

$$P(0,0) = p^4 + p^3q\left(p(1+q) + \frac{p^2q^2}{p^2+q^2}\right)$$
$$+ 3p^2q\left(p^2(1+2q) + \frac{3p^3q^2}{p^2+q^2}\right)$$
$$+ 3p^2q^2\left(p^2 + \frac{2p^3q}{p^2+q^2}\right) + 4p^3q^3\left(\frac{p^2}{p^2+q^2}\right)$$
$$= p^4(1+4q+10q^2) + \frac{20p^5q^3}{p^2+q^2},$$

and the final expression needed is

$$\boxed{P(0,0) = p^4(1+4q+10q^2) + \frac{20p^5q^3}{p^2+q^2}}$$

Plots of the three probabilities, shown in figures 1.3–1.5, for all values of p (remembering that $q = 1 - p$) show that they have very similar behaviour to one another, but there are intersections and if we plot the pairs $\{P(0,0), P(30,15)\}$ and $\{P(0,0), P(40,30)\}$ on the same axes for large p we can see them. This is accomplished in figures 1.6 and 1.7.

Of course, to find those intersections we need to do some algebra.

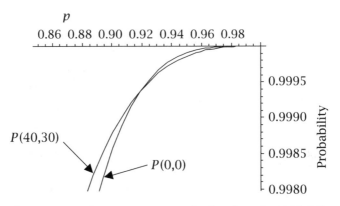

Figure 1.7. The intersection of $P(0,0)$ with $P(40,30)$.

The Intersection of $P(30,15)$ and $P(0,0)$

To find the point of intersection we need to solve the formidable equation

$$p^2(1 + 2q) + \frac{3p^3q^2}{p^2 + q^2} = p^4(1 + 4q + 10q^2) + \frac{20p^5q^3}{p^2 + q^2},$$

again remembering that $q = 1 - p$.

Patience (or good mathematical software) leads to the equation in p,

$$p^2(1 - p)^3(8p^2 - 4p - 3) = 0,$$

which has repeated trivial roots of $p = 0, 1$ as well as the roots of the quadratic equation $8p^2 - 4p - 3 = 0$.

The only positive root is $p = \frac{1}{4}(1 + \sqrt{7}) = 0.911\,437\ldots$ and for any $p > 0.911\,437\ldots$ we will have $P(0,0) > P(30,15)$ and the result for this case is established.

The Intersection of $P(40,30)$ and $P(0,0)$

This time the equation to be solved is

$$p + \frac{p^2q}{p^2 + q^2} = p^4(1 + 4q + 10q^2) + \frac{20p^5q^3}{p^2 + q^2}$$

and, after a similarly extravagant dose of algebra, this reduces to

$$p(1 - p)^3(8p^3 - 4p^2 - 2p - 1) = 0,$$

which again has trivial roots of $p = 0, 1$.

The remaining cubic equation $8p^3 - 4p^2 - 2p - 1 = 0$ has the single real root,

$$p = \tfrac{1}{6} + \tfrac{1}{24}\sqrt[3]{1216 - 192\sqrt{33}} + \tfrac{1}{6}\sqrt[3]{19 + 3\sqrt{33}},$$

which evaluates to $p = 0.919\,643\ldots$.

Again, for any $p > 0.9196\ldots$, we will have $P(0, 0) > P(40, 30)$, with the paradox once again established.

In conclusion, two equal players who are good enough to win the point on their serve just over 90% of the time are better off at the game's start than they are when the score is either 30–15 or 40–30 in their favour.

Chapter 2

THE UPHILL ROLLER

Mechanics is the paradise of the mathematical sciences because by means of it one comes to the fruits of mathematics.

Leonardo da Vinci

An Advertisement for a Book

The Proceedings of the Old Bailey dated 18 April 1694 chronicles a busy day devoted to handing down justice, in which 29 death sentences were passed as well as numerous orders for brandings; there would have been 30 death sentences had not one lady successfully 'pleaded her belly' (that is, proved that she was pregnant). The business part of the document ends with a list of the 29 unfortunates and continues to another list; this time of advertisements (rather strange to the modern mind), which begins with the following paragraph:

> THE Ladies Dictionary: Being a pleasant Entertainment for the Fair Sex; Work never attempted before in English. The Design of this Work is universal, and concerns no less than the whole Sex of Men in some regard, but of Women so perfectly and neatly, that 'twill be serviceable to them in all their Concerns of Business, Life, Houses, Conversations.

Tempting though it is to delve into the details of what suggests itself as a bestselling book, we move to the second advertisement.

> Pleasure with Profit: Consisting of Recreations of divers kinds, viz. Numerical, Geometrical, Mathematical, Astronomical, Arithmetical, Cryptographical, Magnetical, Authentical, Chymical, and Historical. Published to Recreate Ingenious Spirit, and to induce them to make further scrutiny how these (and the like) Sublime Sciences. And to divert them from following such Vices, to which Youth (in this Age) are so much inclin'd. By William Leybourn, Philomathes.

Presumably, those who were tried at the assizes had been given insufficient access to the work and we will touch on only a small part of it ourselves, to be precise, pages 12 and 13.

William Leybourn (1626–1719) (alias Oliver Wallingby) was in his time a distinguished land and quantity surveyor (although he began his working life as a printer). Such was his prestige, he was frequently employed to survey the estates of gentlemen, and he helped to survey the remnants of London after the great fire of 1666. Also, he was a prolific and eclectic author. In 1649 he published (in collaboration with one Vincent Wing) *Urania Practica*, the first book in English devoted to astronomy. After this came *The Compleat Surveyor*, which first appeared in 1653 and ran to five editions, and is regarded as a classic of its kind. His 1667 work, *The Art of Numbering by Speaking Rods: Vulgarly Termed Napier's Bones*, was significant in bringing them further into the public eye.

In 1694 he had published the recreational volume *Pleasure with Profit*, the opening page of which is shown in figure 2.1.

We can readily agree with the following sentiment expressed in the book:

> But leaving those of the Body, I shall proceed to such Recreations as adorn the Mind; of which those of the Mathematicks are inferior to none.

And having done so we can then concentrate on a delightful mechanical puzzle described in the book and attributed to one 'J.P.', which has become known as the Uphill Roller.

Figures 2.2 and 2.3 show pages 12 and 13 of the book, which detail the construction of a double cone and two inclined rails along which the cone can roll – *uphill*. His final paragraph explains the paradox, pointing out that the important issue is that, even though the cone does ascend the slope, its centre of mass will descend if the measurements are just right, which ensures that, although one's senses might be confounded, the law of gravity is not.

An Explanation

Before we examine Leybourn's explanation, we will look at the matter through modern eyes, using elementary trigonometry to study it. Figures 2.4, 2.5 and 2.6 establish the notation that we need and parametrize the configuration in terms of three angles: α, the angle of inclination of the sloping rails; β, the semi-angle between the rails, measured horizontally at floor level; γ, the semi-angle at an apex of the double cone. Write a and b as the heights of the lower and upper ends of the rails and r as the radius of the double cone. An x/y coordinate system is then set up as shown in figure 2.4.

With the coordinate system in place, we can find the equation of the path of the centre of mass of the cone as it rolls up the slope. From figures 2.5 and 2.6 we have

$$PQ = P_1Q_1 = 2x \tan \beta.$$

From figure 2.6 we have

$$RS = PS \tan \gamma = \tfrac{1}{2}PQ \tan \gamma = x \tan \beta \tan \gamma,$$
$$SG = r - RS = r - x \tan \beta \tan \gamma,$$
$$y = PP_1 + SG = PP_1 + (r - x \tan \beta \tan \gamma).$$

From figures 2.4 and 2.6 we have

$$PP_1 = SG_1 = a + x \tan \alpha.$$

Therefore,

$$y = (a + x \tan \alpha) + (r - x \tan \beta \tan \gamma)$$
$$= a + r + x(\tan \alpha - \tan \beta \tan \gamma).$$

The path of the centre of mass of the cone is, then, the straight line

$$y = a + r + x(\tan \alpha - \tan \beta \tan \gamma),$$

which has gradient $\tan \alpha - \tan \beta \tan \gamma$ and for the motion to be possible this gradient has to be negative, which means that the defining condition for the paradox to exist is that

$$\tan \alpha < \tan \beta \tan \gamma.$$

Of course, to appreciate the paradox properly, a physical model is needed and the reader is strongly encouraged to make one (or to get someone else to). The author's model (made by his long-term friend, Brian Caswell) has

$$\alpha = 4.6°, \qquad \beta = 15.3°, \qquad \gamma = 25.4°,$$

from which it is plain that the inequality just holds.

Leybourn's Version

Now that we have a clear criterion to use, we can look more closely at Leybourn's instructions. If we take the diameter of his double cone to be the upper limit of 6 inches and realize that 1 yard is 36 inches, his description translates in our terms to:

$$r = 3,$$
the semi-length of the cone equals $3 \times 3 = 9$,
$$b - a \approx r = 3,$$
and the length of the slope equals 36.

From this we can deduce from figure 2.4 that $\sin \alpha = \frac{3}{36} = \frac{1}{12}$. The lengths of the horizontal projections of the slopes in figure 2.4 are each $36 \cos \alpha$, which makes

$$\sin \beta = \frac{9}{36 \cos \alpha} = \frac{1}{4 \cos \alpha}.$$

From figure 2.6, $\tan \gamma = \frac{3}{9} = \frac{1}{3}$.

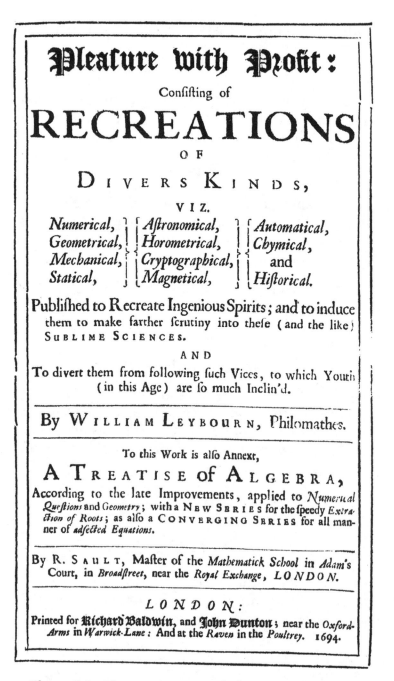

Figure 2.1. The opening page of *Pleasure with Profit*.

RECREATIONS

A
MECHANICAL PARADOX:
O R, A
New and Diverting Experiment.

Whereby a Heavy Body ſhall by its own
Weight move up a ſloping Aſcent.

Written by J. P.

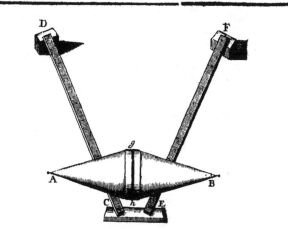

T H E Things neceſſary for this Experiment are, Firſt, A Roller
of Wood, turned in a *Lathe-Ink*, a Figure like as here is re-
preſented at A B, *(viz.)* of two Cones (or Sugar Loaves) abutting.
one againſt the other. Let the thickneſs in the middle (g h) be about
5 or

Figure 2.2. Page 12 of *Pleasure with Profit.*

MECHANICAL.

5 or 6 Inches, the length A B about 3 times the thicknefs; at the end A & B may be left two little Pins turned. 2. You muft provide two ftraight fmooth Rulars about a Yard in length, and ftrong enough to bear the weight of the Roller. 3. Laftly, You muft have three pieces of Wood to fupport the ends of the Rulars; the firft about two or three Inches thick, the other two (to ftand at D and F) muft be thicker than this firft by fomewhat lefs than half the Diameter of the Roller; fo that if the Roller be 6 Inches Diameter, the firft piece of Wood 2 Inches, then let the other 2 pieces be about 4 ¼ Inches apiece. Being thus provided when you would try the Experiment, 1. Place the two thicker pieces upon a level Table almoft the length of the Roller off of one another, as at D F, fet the other piece of Wood almoft the length of the Rular off of the other two. 2. Place the two Rulars with their ends upon the pieces of Wood in the manner as is reprefented in the Figure, with their lower ends near together, and the upper ends ftradling. 3. Place the middle of your Roller between the two lower ends of the Rulars, and you will fee (if you have placed all right) what you defire, *viz.* The Roller will of it felf climb to the upper ends of the Rulars.

When you would divert any perfon with this Experiment, you may firft put the Rulars Parallel, (or with lower ends as wide as the upper) and let it be feen how faft the Roller will run down the Defcent; which will make it the more ftrange to fee it afterwards climb the fame Afcent, by only bringing the lower ends nearer.

The reafon of this (feeming) Afcent of the Rhomb or Roller, is a real *Defcent* or *Lowering* of its Center of Gravity, for tho' the way or line of the motion on the *Rulars* be an *Afcent*, yet the line which the Roller defcribes on its own furface is fuch, that every point of it approaches nearer to the Axis of the Rhomb, the opening of the Rulars caufing the Contact to be nearer to the fmall ends of the two Cones; and confequently, nearer to their Axis: Whereupon the Axis of the Rhomb is fo much lower at the top of the Rulars, as their Elevation comes fhort of the Semi-diameter of the Rhomb.

Figure 2.3. Page 13 of *Pleasure with Profit.*

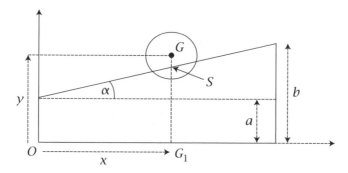

Figure 2.4. A side view, looking at the section along OG_1.

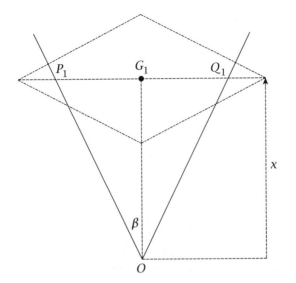

Figure 2.5. A plan view at floor level, with the cone contacting the runners at P and Q, with P_1 and Q_1 lying directly below at floor level.

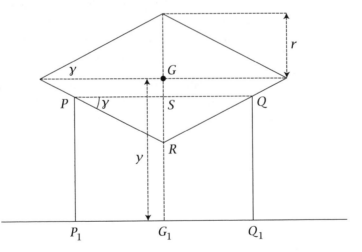

Figure 2.6. A front view, looking from O.

Knowing the exact value of $\sin \alpha$ enables us to use Pythagoras's Theorem to calculate the third side of the right-angled triangle as $\sqrt{12^2 - 1^2} = \sqrt{143}$ and therefore to evaluate $\cos \alpha = \frac{\sqrt{143}}{12}$ and $\tan \alpha = \frac{1}{\sqrt{143}}$.

Now we have that

$$\sin \beta = \frac{1}{4 \cos \alpha} = \frac{3}{\sqrt{143}}$$

and the use of Pythagoras's Theorem once again gives the third side of that triangle to be $\sqrt{143 - 9} = \sqrt{134}$, making $\tan \beta = \frac{3}{\sqrt{134}}$.

In summary, Leybourn's instructions reduce to

$$\tan \alpha = \frac{1}{\sqrt{143}}, \qquad \tan \beta = \frac{3}{\sqrt{134}}, \qquad \tan \gamma = \frac{1}{3},$$

and our inequality requires that $\frac{3}{\sqrt{134}} \times \frac{1}{3} > \frac{1}{\sqrt{143}}$, or that $143 > 134$, which indeed it is! (Using the same analysis, his lower limit of 5 inches requires that $5159 > 4934$.)

With the mystery of the uphill roller explained we will leave the scholarship of Mr Leybourn with part of Samuel Pepys's diary entry of 18 August 1662:

> Mr. Deane of Woolwich came to me, and he and I rid into Waltham Forest, and there we saw many trees of the King's a-hewing; and he showed me the whole mystery of off square wherein the King is abused in the timber that he buys, which I shall with much pleasure be able to correct.

One of Leybourn's publications had explained the fraudulent practice of *off square* cutting to the understanding of Mr Deane, an official at Woolwich.

Chapter 3

THE BIRTHDAY PARADOX

I'm sixty years of age. That's 16 Celsius.

George Carlin

The Basic Problem

Perhaps one of the most well-known examples of a counterintuitive phenomenon concerns the likelihood of two individuals in a gathering sharing the same birthday. If we ignore leap years, then, with a gathering of 366 people, we are assured of at least one repetition of a birthday (a simple application of the subtly powerful Pigeon Hole Principle). That observation is clear enough. What is considerably more perplexing is the size of the group which would result in a 50:50 chance of such a repetition; intuition has commonly argued that, since we halve the probability, we should need about half the number of people, around 183. And intuition is much misguided.

Not the Problem

In part, the counterintuitive nature of the result stems from a common misconception of its statement. It is not that, among r people, at least one person has the same birthday as oneself, although this is easily calculated.

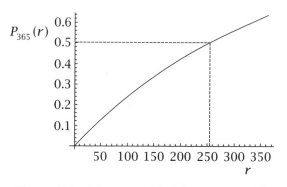

Figure 3.1. The same birthday as oneself.

In an n-day year, the probability that a single person will miss the birthday in question is $(n-1)/n$ and so, assuming independence, the probability that all r of them will do so is $((n-1)/n)^r$, which means that the probability of at least one match is

$$P_n(r) = 1 - \left(\frac{n-1}{n}\right)^r = 1 - \left(1 - \frac{1}{n}\right)^r$$

and so, for a standard year,

$$P_{365}(r) = 1 - \left(\frac{364}{365}\right)^r.$$

It is clear that, as $r \to \infty$, $P_{365}(r) \to 1$ and figure 3.1 shows enough of the plot of $P_{365}(r)$ for the value 0.5 to be exceeded and at a value of r somewhere above 250. To find the exact value of r, we must solve $P_{365}(r) = 0.5$, which means finding r so that $1 - (\frac{364}{365})^r = 0.5$ and so $(\frac{364}{365})^r = 0.5$. This makes

$$r = \frac{\ln 0.5}{\ln(\frac{364}{365})} = 252.65\ldots \approx 253.$$

The correct interpretation of the statement of the problem has us not asking that a particular birthday be matched, but that there exist two birthdays that match. A linear increase in r results in a combinatorial increase in the possible number of matches, by which we mean that as r increases by 1 the number of possible matches obviously increases by r itself. Symbolically,

$$\binom{r+1}{2} - \binom{r}{2} = \frac{(r+1)!}{2!(r+1-2)!} - \frac{r!}{2!(r-2)!}$$

$$= \frac{r!}{2(r-1)!}((r+1) - (r-1)) = r.$$

So, for example, increasing r from 22 to 23 increases the number of possible matches from $\binom{365}{22} = 231$ to $231 + 22 = 253$. It is this significant rate of increase of possible successes that underlies the solution to the problem.

This deals with the common misconception; now we will move to the real thing.

The Standard Solution

The usual analysis for a year of n days and a random collection of r people again uses the standard observation that the probability of at least two people having the same birthday is one minus the probability of everybody having different birthdays. That said, the first person's birthday can be chosen in n out of the n possibilities and, with this one day used up, the second person's birthday can then be chosen in $(n-1)$ ways, etc. Continuing the argument for all r we arrive at the expression for the new $P_n(r)$, the probability that at least two people have the same birthday, as shown below:

$$P_n(r) = 1 - \frac{n}{n} \times \frac{n-1}{n} \times \frac{n-2}{n} \times \cdots \times \frac{n-(r-1)}{n}$$

$$= 1 - \frac{n!}{n^r(n-r)!}$$

$$= 1 - \frac{r!}{n^r}\binom{n}{r}.$$

This means that, for a year of 365 days,

$$P_{365}(r) = 1 - \frac{r!}{365^r}\binom{365}{r}.$$

The graph of this function for r up to 100 is shown in figure 3.2.

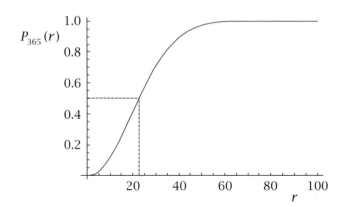

Figure 3.2. The probability of at least two coincident birthdays.

Table 3.1. The critical region.

r	$P_{365}(r)$
21	0.443 688
22	0.475 695
23	0.507 297
24	0.538 344
25	0.568 700

The horizontal line drawn at 0.5 causes us to look at a value of r a little over 20 and table 3.1 details the values in this region and, sure enough, 23 is the critical value for r. To the surprise of many and the shock of some it requires only 23 people to be gathered together for the odds to be in favour of at least two of them sharing a birthday.

Putting the result into a tangible context, in each (English) football match of 11 players a side (plus a referee) the odds are in favour of two of the participants sharing the same birthday. Science journalist Robert Matthews provided some data to support the theory by choosing ten Premiership matches played on 19 April 1996 and establishing birthdays; the results are shown in table 3.2.

With a probability of success of about 0.51 theoretically we would have expected about 5 successes out of the 10 possible matches and we see that there were 6; not such a bad fit.

Table 3.2. Data from ten premiership football matches.

Match	Coincident birthdays
Arsenal v Blackburn	No coincidences
Aston Villa v Tottenham	Ehiogu (A; 3.11.72) and Yorke (A; 3.11.71)
Chelsea v Leicester City	Petrescu (C; 22.12.67) and Morris (C; 22.12.78) Hughes (C; 1.11.63) and Elliott (L; 1.11.68)
Liverpool v Manchester United	James (L; 1.8.70) and Wright (L; 1.8.63) Butt (M; 21.1.75) and P. Neville (M; 21.1.77)
Middlesborough v Sunderland	Johnston (S; 14.12.73) and Waddle (S; 14.12.60)
Newcastle v Derby	No coincidences
Nottingham Forest v Leeds	Martyn (L; 11.8.66) and Halle (L; 11.8.65)
Sheffield Wednesday v Wimbledon	No coincidences
Southampton v Coventry	Benali (S; 30.12.68) and Whelan (C; 30.12.74)
West Ham v Everton	No coincidences

Assumptions

Throughout, we have assumed that birthdays are evenly distributed throughout the year, which is convenient for our calculations but not strictly true. That said, it has been shown that (not unreasonably) nonuniformity increases the probability of a shared birthday (see, for example, D. M. Bloom (1973), A birthday problem, *American Mathematical Monthly* 80:1141–42, and A. G. Munford (1977), A note on the uniformity assumption in the birthday problem, *American Statistician* 31:119). T. Knapp examined the implications from an empirical viewpoint in his 1982 article, The birthday problem: some empirical data and some approximations, *Teaching Statistics* 4(1):10–14. The empirical data were culled from birth-date information from

Table 3.3. Multiply shared birthdays.

n	r	n	r
2	23	9	985
3	88	10	1181
4	187	11	1385
5	313	12	1596
6	460	13	1813
7	623	14	2035
8	798	15	2263

Monroe County, New York, over the 28-year-period 1941–1968 (the length of the cycle chosen to smooth out micro fluctuations): the discrepancy was minuscule.

What difference does a leap year make? Again, not very much. If we model the situation using a year of 365.25 days with the assumption that the probability of being born on 29 February is 0.25 of that on any other day, we have that the probability of a randomly selected person being born on 29 February is 0.25/365.25, and the probability that a randomly selected person was born on another specified day is 1/365.25. More (slightly more delicate) calculations reveal that 23 is again the magic number with the only difference that the associated probability is 0.5068....

Generalization

There are simple ways of generalizing the problem: for example, we might ask how many people are needed for the odds to be in favour of at least two of them being born in the same month, or having the same birth sign. Putting $n = 12$ in the formula for $P_n(r)$ reveals that $r = 4$ gives the probability as 0.427 083... and $r = 5$ gives it as 0.618 056....

A question which is much harder to solve is to find the minimum number of people, r, for which the odds are in favour of at least $3, 4, \ldots, n$ of them sharing the same birthday. R. J. McGregor and G. P. Shannon (for example) gave such an analysis using the theory of partitions in their 2004 paper, On the generalized birthday problem, *Mathematical Gazette* 88(512):242–48. The first few values of n and r are given in table 3.3.

Finally, we might ask the probability that, among r people and with a year of 365 days, there is a 'near-miss' of birthdays. To be exact, we ask to calculate the probability

$$P_r(\geqslant 2 \text{ birthdays separated by} \leqslant d \text{ days}).$$

Again, this is quite a difficult problem (see J. I. Naus (1968), An extension of the birthday problem, *American Statistician* 22:27–29). His calculations reveal that

$P_r(\geqslant 2$ birthdays separated by $\leqslant d$ days)

$$= 1 - \left[\left(3651(364 - rd)! - \frac{r}{(365 - (d+1)r!)}\right)\right].$$

Matthews calculated this probability for birthdays either on the same day or on adjacent days (taking $d = 1$) for the football example (taking $r = 23$) to get the value 0.888.... This means that we would expect about 9 of the 10 fixtures to possess this attribute; using his complete dataset he points out that, in fact, all 10 do.

Finally, this last formula can be used to calculate the minimum r for which

$$P_r(\geqslant 2 \text{ birthdays separated by} \leqslant d \text{ days}) \approx 0.5$$

for any values of d. Table 3.4 shows the results of calculating this probability for d between 0 and 7, with the first row of data reflecting the Birthday Paradox. The last row is rather surprising too; it says that in a family of six members it is more than likely that two of them will have a birthday within a week of each other.

Halmos's Answer

The late Paul Halmos, who wrote, taught and inspired for decades, is quoted as saying that 'computers are important, but not to mathematics'. In particular, in his autobiography, *I Want*

Table 3.4. Birthdays separated by up to a week.

d	r
0	23
1	14
2	11
3	9
4	8
5	7
7	6

to Be a Mathematician, he deplored the fact that the Birthday Paradox is customarily solved by a computational method, for example, as shown in the section on the standard answer earlier in the chapter. He expressed the view that it was naturally susceptible to analysis and provided the following argument to justify the claim. The method also gives a useful asymptotic estimate of the probability for large n. It is also very pretty.

He stated in that autobiography that

> A good way to attack the problem is to pose it in reverse: what's the largest number of people for which the probability is less than 1/2 that they all have different birthdays?

In terms of our original notation this means that, for a given n, we require the largest r such that

$$\frac{n}{n} \times \frac{n-1}{n} \times \frac{n-2}{n} \times \cdots \times \frac{n-(r-1)}{n} < \frac{1}{2}$$

or

$$1 \times \left(1 - \frac{1}{n}\right) \times \left(1 - \frac{2}{n}\right) \times \cdots \times \left(1 - \frac{r-1}{n}\right) < \frac{1}{2}.$$

The next step is to invoke the arithmetic geometric mean inequality, which states that, for any r positive numbers,

$$\sqrt[r]{a_1 a_2 a_3 \cdots a_r} \leqslant \frac{a_1 + a_2 + a_3 + \cdots + a_r}{r},$$

with equality only when all numbers are equal.

With

$$a_k = \left(1 - \frac{k-1}{n}\right) \quad \text{for } k = 1, 2, 3, \ldots, r$$

we then have that

$$\sqrt[r]{1 \times \left(1 - \frac{1}{n}\right) \times \left(1 - \frac{2}{n}\right) \times \cdots \times \left(1 - \frac{r-1}{n}\right)}$$

$$\leqslant \frac{1 + \left(1 - \frac{1}{n}\right) + \left(1 - \frac{2}{n}\right) + \cdots + \left(1 - \frac{r-1}{n}\right)}{r}$$

$$= \frac{1}{r} \sum_{k=0}^{r-1} \left(1 - \frac{k}{n}\right)$$

$$= \frac{1}{r} \left(\sum_{k=0}^{r-1} 1 - \sum_{k=0}^{r-1} \frac{k}{n}\right)$$

$$= \frac{1}{r} \left(r - \frac{1}{n} \times \frac{r-1}{2} \times r\right)$$

$$= \left(1 - \frac{r-1}{2n}\right) \approx \left(1 - \frac{r}{2n}\right)$$

and so,

$$1 \times \left(1 - \frac{1}{n}\right) \times \left(1 - \frac{2}{n}\right) \times \cdots \times \left(1 - \frac{r-1}{n}\right) \leqslant \left(1 - \frac{r}{2n}\right)^r.$$

Next we invoke the inequality $1 - x \leqslant e^{-x}$ for $x \geqslant 0$ to get $1 - r/2n \leqslant e^{-r/2n}$ and this means that

$$1 \times \left(1 - \frac{1}{n}\right) \times \left(1 - \frac{2}{n}\right) \times \cdots \times \left(1 - \frac{r-1}{n}\right)$$

$$\leqslant \left(1 - \frac{r}{2n}\right)^r \leqslant (e^{-r/2n})^r = e^{-r^2/2n}.$$

Finding the smallest r such that $e^{-r^2/2n} \leqslant \frac{1}{2}$ will then give us an upper bound on the smallest r such that

$$1 \times \left(1 - \frac{1}{n}\right) \times \left(1 - \frac{2}{n}\right) \times \cdots \times \left(1 - \frac{r-1}{n}\right) < \frac{1}{2}.$$

If we consider the 'equation' $e^{-r^2/2n} \approx \frac{1}{2}$ and take natural logs of both sides, we have that

$$-\frac{r^2}{2n} \approx -\ln 2 \quad \text{and} \quad r \approx \sqrt{2\ln 2}\sqrt{n} \approx 1.18\sqrt{n}.$$

So, for a group of a 'large' size n we need a sample of at most $1.18\sqrt{n}$ to have an even chance of two of them matching and we can quantify the surprise in the Birthday Paradox by stating that the minimum number required to fulfil the requirements is of the order $\sqrt{n}$. With all of this hand-waving approximation around it is comforting to check that the formula for $n = 365$ gives the very accurate estimate 22.54.

Halmos continued by saying that

> The reasoning is based on important tools that all students of mathematics should have ready access to. The birthday problem used to be a splendid illustration of the advantages of pure thought over mechanical manipulation; the inequalities can be obtained in a minute or two, whereas the multiplications would take much longer, and be much more subject to error, whether the instrument is a pencil or an old-fashioned desk computer. What calculators do not yield is understanding, or mathematical facility, or a solid basis for more advanced, generalized theories.

Also, calculators (and computers) eventually cannot cope with very large numbers and this result provides the mechanism for doing just that, as we shall now see.

A Practical Application

The Birthday Paradox is more than a novelty; in fact, it has applications in many areas, including cryptography, sorting and the somewhat esoteric code numbers called GUIDs, which identify the product of a particular package of a particular computer. Using the estimate derived from Halmos's ideas, we can look at these Globally Unique Identifiers (GUIDs).

Each GUID is 128 bits long, which means that it can be written as a 32 digit hexadecimal (base 16) number. There are many

Internet sites which will provide such a number using one algorithm or another and, in particular, one site contained the following:

> GUID.org is an Internet service that assigns anonymous user IDs to web browsers. These anonymous IDs can then be used by other web sites for many purposes. For example, a site may use your GUID to recognize you when you return.
>
> GUID.org works by assigning each browser a unique, essentially random 16-byte user ID, which is represented as 32 hexadecimal digits. This ID is constructed by applying a MD5 hash to a string concatenated from the IP address of the requestor, the IP address of this server, the date, and the time of day in ticks. The ID is then set as a cookie from GUID.org.

Never mind about the computer jargon, the important matter is that a GUID is randomly generated, 128 bits long and supposedly unique, and it may be important that it is unique if it is to be used to identify a revisit to a site. It happens that, when the author requested a GUID from the site, the GUID assigned was

B46F DD75 A69B 809F 3A16 636C C892 116F

using the standard digits $\{1, 2, 3, \ldots, 9, A, B, \ldots, F\}$ of the hexadecimal number system.

Could the number be safely used as a unique identity code? There is a chance that this random process will result in a GUID which has been used before; but what chance? This is just a disguised form of the birthday problem with $n = 2^{128}$. Using the above result we can see that the total number of GUIDs generated before the odds are in favour of a clash is about $1.18 \times \sqrt{2^{128}} = 1.18 \times 2^{64} \approx 2.18 \times 10^{19}$, which is vastly big.

To give an idea of the size of things, if 100 000 GUIDs were being generated every hour of every day it would take about 22 billion years to generate this number – and the universe is only about 12–15 billion years old. The system seems reasonably safe!

As a final hint as to the generality of the application of the result, the reader might wish to pursue the following item:

M. H. Gail, G. H. Weiss, N. Mantel and S. J. O'Brien (1979), A solution to the generalized birthday problem with application to allozyme screening for cell culture contamination, *Journal of Applied Probability* 16:242-51.

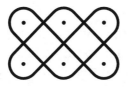

Chapter 4

THE SPIN OF A TABLE

In mathematics, you don't understand things. You just get used to them.

John von Neumann

The Original Problem

Martin Gardner brought to the wider world 'a delightful combinatorial problem of unknown origin' in his February 1979 column in *Scientific American*. He commented that Robert Tappay of Toronto had passed it to him, who believed it to have originated in Russia:

> A square table stands on a central column, which allows it to rotate freely in a horizontal plane. At each corner there is a pocket too deep to allow the contents to be seen and of a size to accommodate an ordinary, empty wine glass. An electronic mechanism is fitted so that, with each pocket containing a single wine glass, a bell will ring if all glasses are oriented in the same direction. The experiment begins with the glasses distributed between the pockets, their orientation randomly chosen. A person sits at the table and chooses two pockets simultaneously, from which the glasses are removed, examined and replaced as that person decides.

The table is then spun in such a way that the person is unable to tell which side now is in front. The process is then repeated indefinitely.

After any repetition the bell might or might not ring simply by chance but the problem is to find a procedure which will ensure that it does ring after a finite number of spins. This is not a case of probability, not a matter of arguing that eventually the bell *must surely* ring: it will ring with absolute certainty.

Two Simpler Cases

As is so often the case, it helps to illuminate matters if we consider simpler cases, in this case a table with just two pockets, modelled by a rod with pockets at either end, or three pockets, modelled by an equilateral triangle with pockets at each vertex. Figure 4.1 provides diagrams of such tables.

The first thing to realize is that we can assume that the initial random placement has the glasses in different orientations – otherwise the bell would ring straight away.

With two pockets the problem is entirely trivial: since the bell does not ring when the glasses are put in the pockets, when the table has stopped spinning look at both glasses and invert one of them to ensure that they both have the same orientation.

Now suppose that the table is in the shape of an equilateral triangle, with a pocket at each vertex. The following procedure will guarantee that the bell rings:

(1) Reach into any pair of pockets; if the glasses are oriented in the same direction invert them both and the bell will ring. Otherwise the glasses will be facing in different directions, so invert the glass that is facing down.

(2) If the bell does not ring, spin the table and reach into any two pockets; if both glasses are turned up, invert both and the bell will ring. If they are turned in opposite directions, invert the glass turned down and the bell will ring.

From these two simple cases we can see that the result for four pockets is at least plausible and we are now ready to consider this original puzzle.

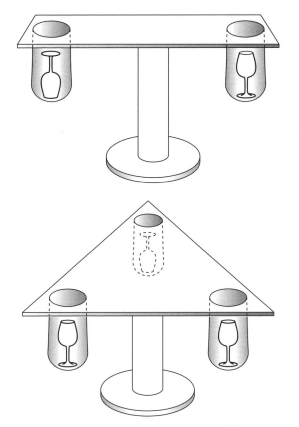

Figure 4.1. Two simpler situations.

The Original Problem Solved

Figure 4.2 shows our new table. An initial, important observation for this case is that the selection of the pockets has essentially two forms: a side pair or a diagonal pair. It is also clear that these choices must alternate, otherwise we could go on repeating ourselves forever. With that in mind we can look at a procedure which guarantees that the bell will ring.

(1) Reach into a diagonal pair of pockets and orient the glasses to be the same way up.

(2) Given that the bell does not ring, spin the table and reach into two adjacent pockets. If the glasses are both turned up, leave them, otherwise invert the glass that is turned down.

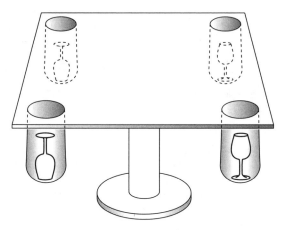

Figure 4.2. The square table.

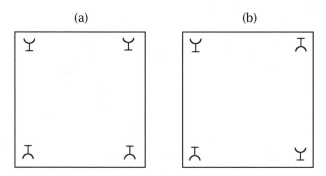

Figure 4.3. The two assured orientations.

If the bell does not ring, it is certain that there are three glasses with the same orientation.

(3) Spin the table and then reach into a diagonal pair of pockets. If one of the glasses is turned down, invert it and the bell will ring. If both are turned up, invert one of them, in which case the orientations will be as in figure 4.3(a).

(4) Spin the table, reach into two adjacent pockets and invert the glasses. If they were both of the same orientation, the bell will ring, otherwise the glasses will now be as in figure 4.3(b).

(5) Spin the table, reach into a diagonal pair of pockets and invert both glasses. The bell will definitely ring.

With this argument, the problem is solved in at most five spins of the table (which is minimal). If we decide to sacrifice minimalism (and thought), the following seven steps solve the problem automatically:

(1) Invert any diagonal pair.
(2) Invert any adjacent pair.
(3) Invert any diagonal pair.
(4) Invert any single glass.
(5) Invert any diagonal pair.
(6) Invert any adjacent pair.
(7) Invert any diagonal pair.

The Problem Generalized

In the end, a table with two pockets presents a trivial problem, one with three pockets an easy problem and one with four pockets a rather more subtle problem. What about a table with five pockets? The answer is that the situation changes radically since, with a five-sided (or greater) table, there is no algorithm which will guarantee that the bell will ring in a finite number of moves. (In chapter 6 we will meet a second situation in which matters change radically at the fifth level, a phenomenon which is far from uncommon in mathematics.)

The March 1979 *Scientific American* column provided the solution to the original problem. Evidently, mathematicians had been active between the February and March issues, since the March column also mentions two generalizations suggested by Ronald L. Graham and Persi Diaconis:

(1) Can the bell be made to ring if the player is replaced by an 'octopus' with k hands sitting at a table with n sides?
(2) Can the bell be made to ring if the glasses are replaced by objects which can occupy more than two positions?

They provided a partial solution to the first question, showing that with a table having a prime number of sides n the minimum number of hands needed to guarantee the bell ringing is $n-1$ and that the minimum number is bounded above by $n-2$ otherwise.

Of course, their result decides the case for the five-sided table mentioned above; there, $k = 2$ and $n = 5$.

Subsequently, William T. Laaser and Lyle Ramshaw, both of Stanford, solved the first generalization completely. Their result is that the minimum number of hands, k, required to ensure that the bell will ring for an n-sided table is $k = (1 - 1/p)n$, where p is the largest prime factor of n (a formula conjectured by James Boyce). Of course, this reduces to the above result in the case where n is prime (and therefore $n = p$).

The full exposition of the Laaser–Ramshaw result (Probing the rotating table, *Mathematical Gardner*, 1981, 288-307) is too long for inclusion here but we will consider the first part of it, which establishes that $(1 - 1/p)n$ is a lower bound on k; that is, if $k < (1 - 1/p)n$, it is impossible to guarantee that the bell will ring.

First we will establish a preliminary result.

Consider the set of integers $\{0, 1, 2, \ldots, p-1\}$ reduced modulo the prime p. If we start at any position r and move through the integers in steps of 1 (reducing modulo p), it is evident that we will visit each integer before we reach our starting point again. Now suppose that we move in steps of size j (where $2 \leqslant j \leqslant p - 1$). We will generate the set of integers $\{r + \alpha j : 0 \leqslant \alpha \leqslant p - 1\}$, modulo p, as we move through the integers and if two of these numbers are equal it must be that $r + \alpha j = r + \beta j$, modulo p, and this means that $(\beta - \alpha)j$ is divisible by p. Since p is prime and cannot possibly divide j, it must be that p divides $\beta - \alpha$ and this makes $\beta = \alpha + Np$. In short, any walk around the integers will visit each one of them once before returning to the starting point.

With that in place consider the Laaser-Ramshaw result in two parts:

(1) Suppose that $n = p$ is prime. If the player has fewer than $p - 1$ hands, then he has no winning strategy.

If the player has fewer than $p - 1$ hands, any probe of the table will leave at least two pockets untested; call these pockets *gaps* and suppose that two of them are a distance j apart. Our

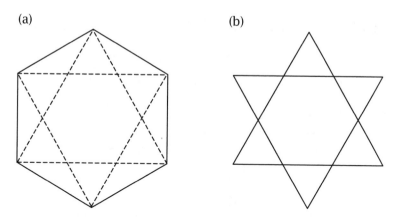

Figure 4.4. A table and its sub-tables.

preliminary argument shows that, if we start at any pocket and walk around the table in steps of length j, we will visit every pocket before returning to the starting point. Since the pockets must contain glasses of both orientations, it must be that our journey will at some stage cause us to step from a glass of one orientation to one of another, precisely a distance j apart. If the table happens to align itself so that the gaps j apart in the probe pattern match the glasses of different orientation, the bell cannot ring. The procedure can be repeated indefinitely, which shows that no ringing of the bell can be guaranteed.

(2) Now suppose that $n \geqslant 2$ is composite and let p be its largest prime factor. If the player has fewer than $(1 - 1/p)n$ hands, then he has no winning strategy.

Write $n = pl$. The argument essentially reduces this case to the previous one. Rather than consider the table cyclically, picture it as l copies of a sub-table of p sides. For example, if $n = 6 = 2 \times 3$, we can think of the hexagonal table in figure 4.4(a) as the superposition of the two triangular sub-tables in figure 4.4(b).

Since the player has fewer than

$$\left(1 - \frac{1}{p}\right)n = \left(1 - \frac{1}{p}\right)pl = (p - 1)l$$

hands at any probe, at least one of the sub-tables will have at least two gaps when the full table is probed, since if each had at

Table 4.1. Minimum number of hands, N, needed for n-sided tables.

n	N	n	N	n	N	n	N
4	2	20	16	36	24	52	48
5	4	21	18	37	36	53	52
6	4	22	20	38	36	54	36
7	6	23	22	39	36	55	50
8	4	24	16	40	32	56	48
9	6	25	20	41	40	57	54
10	8	26	24	42	36	58	56
11	10	27	18	43	42	59	58
12	8	28	24	44	40	60	48
13	12	29	28	45	36	61	60
14	12	30	24	46	44	62	60
15	12	31	30	47	46	63	54
16	8	32	16	48	32	64	32
17	16	33	30	49	42	65	60
18	12	34	32	50	40	66	60
19	18	35	30	51	48	67	66

most one gap, there would have to be at least $l \times (p-1) = (p-1)l$ hands. Suppose that on a sub-table the two gaps are a distance j apart.

There must exist a sub-table whose pockets contain glasses of both orientations. Take a walk around it as before in steps of length j and, as before, every pocket will be visited before returning to the starting place and this means that there will be two pockets, a distance j apart, one of which contains an 'up' glass and the other a 'down' glass. If the table happens to align itself so that a probe's sub-table with two gaps is aligned with the sub-table with the glasses of both orientations and with the gaps and the up and down glasses superimposed, the bell cannot ring. This can continue indefinitely, denying any possibility of an assured ringing of the bell.

We should note in passing that (for $n > 2$), since

$$\left(1 - \frac{1}{p}\right)n = \frac{p-1}{p}n,$$

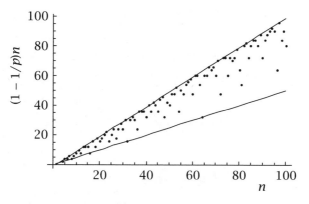

Figure 4.5.

if $p = 2$, then n must be a power of 2, otherwise $p > 2$ and so $p - 1$ is even; either way (rather strangely), $(1 - 1/p)n$ is even.

We will finish with table 4.1, which gives the values of $N = (1 - 1/p)n$ for the first few values of n, and figure 4.5, which graphs the values for n up to 100. The trend is clear and reasonable, but regularly upset (most particularly) when n is a power of 2, in which case $p = 2$ and $(1 - 1/p)n = \frac{1}{2}n$.

In fact, it is evident that the following bounds exist for N

$$\tfrac{1}{2}n \leqslant N \leqslant n - 1,$$

with the lower limit achieved when N is a power of 2 and the upper limit when N is prime. This is indicated by the lines $y = \frac{1}{2}n$ and $y = n - 1$, which have been added to the plot.

Chapter 5

DERANGEMENTS

I think it is said that Gauss had ten different proofs for the
law of quadratic reciprocity. Any good theorem should have
several proofs, the more the better. For two reasons: usually,
different proofs have different strengths and weaknesses,
and they generalise in different directions: they are not just
repetitions of each other.

Sir Michael Atiyah

We shall look at a famous old problem in three different, enlight-
ening ways and then consider three surprising facts originating
from it.

An Old Card Game

The French word for 13, *trieze*, was also the name of a commonly played card game of the eighteenth century. It could be
considered as a simple patience (or solitaire) game but in its classic form it was played by several individuals, and commonly for
money. We will leave it to the man who is credited with its first
analysis to explain matters:

> The players draw first for who will have the hand. We sup-
> pose that this is Pierre, & that the number of the players

is as such as one would wish. Pierre having an entire deck composed of fifty-two cards shuffled at discretion, draws them one after the other, naming & pronouncing one when he draws the first card, two when he draws the second, three when he draws the third, & thus in sequence up to the thirteenth which is a King. Now if in all this sequence of cards he has drawn none of them according to the rank that he has named them, he pays that which each of the players has wagered in the game, & gives the hand to the one who follows him at the right. But if it happens to him in the sequence of thirteen cards, to draw the card which he names, for example, to draw an ace at the time which he names one, or a two at the time which he names two, or a three at the time which he names three, &c. he takes all that which is in the game, & restarts as before, naming one, next two, &c. It is able to happen that Pierre having won many times, & restarting with one, has not enough cards in his hand in order to go up to thirteen, now he must, when the deck falls short to him, to shuffle the cards, to give to cut, & next to draw from the entire deck the number of cards which is necessary to him in order to continue the game, by commencing with the one where he is stopped in the preceding hand. For example, if drawing the last card from them he has named seven, he must in drawing the first card from the entire deck, after one has cut, to name eight, & next nine, &c. up to thirteen, unless he rather not win, in which case he would restart, naming first one, next two, & the rest as it happens in the explanation. Whence it seems that Pierre is able to make many hands in sequence, & likewise he is able to continue the game indefinitely.

This extract (for which we have relied on the translation by Richard J. Pulskamp) is from the book *Essai d'analyse sur les jeux de hazard*, 2nd edn (1713), by Pierre Renard de Montmort. It is followed by an analysis which, in true mathematical fashion, starts with easier cases, moving to a full solution (with significant contributions from Nicolas Bernoulli). Subsequently other luminaries considered variations of the problem, including De Moivre, Euler, Lambert and Laplace, and we will consider what is probably its most common modern form with its most common name of *rencontre* (a French word which can be translated as 'meet by chance'), in which the thirteen card limit is replaced by the whole fifty-two cards of the pack. Since it is the form in which Euler considered the problem we will let him explain it with this extract from the article, Calcul de la probabilité dans le jeu de rencontre, which appears in *Memoires de l'academie des sciences de Berlin* 7, 1753 (again we have relied on the translation of Richard J. Pulskamp):

> The game of rencontre is a game of chance where two persons, each having an entire deck of cards, draw from it at the same time one card after the other, until it happens that they encounter the same card; and then one of the two persons wins. Now, when such an encounter does not happen at all, then it is the other of the two persons who wins. This posed, one asks the probability that each of these two persons will win.

Whatever the order of the cards in one pack, the other pack will consist of some permutation of those cards and we can look at the game from one of the players' points of view by considering the chance that no *encounter* occurs, which brings us to a useful definition.

Derangements

A permutation which leaves no element fixed has become known as a *derangement*; put another way, a derangement of n objects is a permutation of them without fixed points. The number of derangements of n objects is usually written as $!n$, spoken as subfactorial n.

Of course, not all permutations are derangements: for example, $\{5, 1, 2, 3, 4\}$ is a permutation of $\{1, 2, 3, 4, 5\}$ in which no number occupies its original place and so is a derangement, yet $\{5, 2, 1, 3, 4\}$ is not, as the 2 remains fixed.

If we take Montmort's approach and look at the three simplest cases, we can easily see that the single permutation of $\{1\}$ has no derangements, those of $\{1, 2\}$ have the single derangement

$\{2,1\}$ and those of $\{1,2,3\}$ have the two derangements $\{2,3,1\}$ and $\{3,1,2\}$; this means that

$$!1 = 0, \qquad !2 = 1, \qquad !3 = 2.$$

In general, we need to ask the question: Of the $!n$ different permutations of n distinct objects, how many leave no object in its original place?

The answer to the question will provide a general formula for $!n$ and so for $p_n = !n/n!$, the probability that a permutation of n objects is a derangement.

Before we begin, we have said that the notation $!n$ is standard, but it will be convenient for us to use the alternative $D_n = !n$. The use of $!n$ does have something of a visual disadvantage when expressions contain both factorials and subfactorials, as some will in what follows; also, not that it matters to us, the expression $!n!$ is ambiguous. For example, $!3!$ might mean $(!3)! = 2! = 2$ or $!(3!) = !6 = 265$; compare this with the respective equivalents of $D_3! = 2$ and $D_{3!} = 265$.

A First Solution

First, we will find a recurrence relation for D_n.

If $\{a_1, a_2, a_3, \ldots, a_n\}$ is a derangement of $\{1,2,3,\ldots,n\}$, it must be that $a_1 \neq 1$, which leaves $n - 1$ possibilities for it; for the sake of illustration we will assume that $a_1 = 2$. Now let d_n be the number of such derangements, which means that $D_n = (n-1)d_n$. Now there are two possibilities:

(1) $a_2 = 1$, which means that the derangement has the form $\{2, 1, a_3, a_4, a_5, \ldots, a_n\}$, where $\{a_3, a_4, a_5, \ldots, a_n\}$ is a derangement of $\{3, 4, 5, \ldots, n\}$, and there are exactly D_{n-2} of these;

(2) $a_2 \neq 1$, with $\{a_2, a_3, a_4, \ldots, a_n\}$ a derangement of $\{1, 3, 4, \ldots, n\}$; there are D_{n-1} of these.

These combine to mean that $d_n = D_{n-1} + D_{n-2}$ and therefore

$$D_n = (n-1)(D_{n-1} + D_{n-2}), \quad n \geqslant 3.$$

(Incidentally, this means that D_n must be divisible by $n - 1$.)

Table 5.1. Numbers of derangements for small sets.

n	D_n
1	0
2	1
3	2
4	9
5	44
6	265

Knowing that $D_1 = 0$ and $D_2 = 1$, this relation will allow us to generate D_n for any n and table 5.1 shows the first few of them.

(Using this result and induction, it is also easy to establish that $D_n = nD_{n-1} + (-1)^n$.)

We are interested in $p_n = D_n/n!$, the probability of a permutation of n objects being a derangement, and the recurrence relation that we have just derived allows us to begin to find a general expression for this:

$$
\begin{aligned}
p_n &= \frac{D_n}{n!} = \frac{(n-1)(D_{n-1} + D_{n-2})}{n!} \\
&= (n-1)\left[\frac{1}{n}\frac{D_{n-1}}{(n-1)!} + \frac{1}{n(n-1)}\frac{D_{n-2}}{(n-2)!}\right] \\
&= (n-1)\left[\frac{1}{n}p_{n-1} + \frac{1}{n(n-1)}p_{n-2}\right] \\
&= \left(1 - \frac{1}{n}\right)p_{n-1} + \frac{1}{n}p_{n-2} \\
&= p_{n-1} - \frac{1}{n}(p_{n-1} - p_{n-2}).
\end{aligned}
$$

And so,

$$
p_n - p_{n-1} = -\frac{1}{n}(p_{n-1} - p_{n-2}).
$$

And we now have a recurrence relation for p_n, which we can most easily deal with by writing $q_n = p_n - p_{n-1}$ and chasing down the expressions to get

$$q_n = -\frac{1}{n}q_{n-1} = -\frac{1}{n}\left(-\frac{1}{n-1}\right)q_{n-2}$$

$$= -\frac{1}{n}\left(-\frac{1}{n-1}\right)\left(-\frac{1}{n-2}\right)q_{n-3}$$

$$= -\frac{1}{n}\left(-\frac{1}{n-1}\right)\left(-\frac{1}{n-2}\right)\left(-\frac{1}{n-3}\right)q_{n-4}\cdots$$

$$= (-1)^{n-2}\frac{1}{n}\left(\frac{1}{n-1}\right)\left(\frac{1}{n-2}\right)\left(\frac{1}{n-3}\right)\cdots\left(\frac{1}{3}\right)q_2,$$

where

$$q_2 = p_2 - p_1 = \frac{D_2}{2!} - \frac{D_1}{1!} = \frac{1}{2} - 0 = \frac{1}{2}.$$

This means that we may write

$$q_n = (-1)^n\frac{1}{n!},$$

tidying up the -1 term.

Now, if we write the expressions for q_n explicitly, we get

$$q_n = p_n - p_{n-1},$$

$$q_{n-1} = p_{n-1} - p_{n-2},$$

$$q_{n-2} = p_{n-2} - p_{n-3},$$

$$\vdots$$

$$q_2 = p_2 - p_1,$$

and if we add the equations vertically we have, after almost complete cancellation on the right-hand side,

$$p_n - p_1 = p_n - 0 = p_n = \sum_{r=2}^{n} q_r = \sum_{r=2}^{n} (-1)^r\frac{1}{r!}$$

$$= \frac{1}{2!} - \frac{1}{3!} + \frac{1}{4!} - \cdots + (-1)^n\frac{1}{n!},$$

which can conveniently be written as

$$p_n = 1 - \frac{1}{1!} + \frac{1}{2!} - \frac{1}{3!} + \frac{1}{4!} - \cdots + (-1)^n \frac{1}{n!}$$

and we have the expression we seek.

Bernoulli's Solution

The more powerful the mathematical tools used to prove a result, the shorter that proof might be expected to be and we should not ignore the significantly shorter attack which is based on the inclusion–exclusion principle. The general principle is discussed in appendix A and the eminent Nicholas Bernoulli used it to establish the formula for D_n in the following way.

Using the inclusion–exclusion principle we have that

D_n = the total number of permutations of $\{1, 2, 3, \ldots, n\}$

$- \sum_{\{i\}}$ the total number of permutations of

$\{1, 2, 3, \ldots, n\}$ which fix i

$+ \sum_{\{i,j\}}$ the total number of permutations of

$\{1, 2, 3, \ldots, n\}$ which fix $\{i, j\}$

$- \sum_{\{i,j,k\}}$ the total number of permutations of

$\{1, 2, 3, \ldots, n\}$ which fix $\{i, j, k\}$ ...

with the series finishing with the number 1, which is the total number of permutations fixing all n numbers. So,

$$D_n = n! - n(n-1)! + \binom{n}{2}(n-2)! - \binom{n}{3}(n-3)! + \cdots 1$$

with the first part of each term the number of ways of choosing the numbers to be fixed and the second the number of permutations of what remains. Simplifying gives

$$D_n = n! - n! + \frac{n!}{2!(n-2)!}(n-2)! - \frac{n!}{3!(n-3)!}(n-3)! + \cdots 1.$$

And so

$$D_n = n!\left(1 - \frac{1}{1!} + \frac{1}{2!} - \frac{1}{3!} + \cdots + (-1)^n \frac{1}{n!}\right)$$

and, once again,

$$p_n = \frac{D_n}{n!} = 1 - \frac{1}{1!} + \frac{1}{2!} - \frac{1}{3!} + \cdots + (-1)^n \frac{1}{n!}$$

The Final Proof

Following Heba Hathout's article 'The old hats problem'[1], we can count the $n!$ ways of arranging the n objects by partitioning the ways into $n + 1$ disjoint subsets $S_0, S_1, S_2, \ldots, S_n$, where S_r is the set of permutations in which there are exactly $n - r$ fixed points, and we will write $N(S_r)$ for the number of elements in S_r. For example, if there are two fixed points, we have the subset of permutations S_{n-2} with the two fixed points chosen in $\binom{n}{2}$ possible ways: this means that

$$N(S_{n-2}) = \binom{n}{2} D_{n-2}.$$

Continuing the argument results in the total number of permutations of the n objects decomposed in terms of the $N(S_r)$ as

$$n! = N(S_0) + N(S_1) + N(S_2) + \cdots + N(S_n)$$
$$= \binom{n}{n} D_n + \binom{n}{n-1} D_{n-1} + \binom{n}{n-2} D_{n-2} + \cdots + \binom{n}{0} D_0$$
$$= \sum_{r=0}^{n} \binom{n}{r} D_r. \tag{1}$$

We then have that

$$n! = \sum_{r=0}^{n} \binom{n}{r} D_r$$

---

[1] Available at www.rose-hulman.edu/mathjournal/archives/2003/vol4-n1/paper2/v4n1-2do.doc.

and this is a special form of an expression amenable to Binomial Inversion, as described in appendix B.

The statement of this result is that, if two sets of numbers

$$\{a_0, a_1, a_2, \ldots, a_n\} \quad \text{and} \quad \{b_0, b_1, b_2, \ldots, b_n\}$$

are related by the condition

$$b_n = \sum_{r=0}^{n} \binom{n}{r} a_r,$$

then

$$a_n = \sum_{r=0}^{n} (-1)^{n-r} \binom{n}{r} b_r.$$

The result makes the a_r rather than the b_r the subject of the formula.

In our case, writing $b_n = n!$ and $a_r = D_r$ means that we have our

$$b_n = \sum_{r=0}^{n} \binom{n}{r} a_r$$

and so

$$a_n = \sum_{r=0}^{n} (-1)^{n-r} \binom{n}{r} b_r$$

becomes

$$D_n = \sum_{r=0}^{n} (-1)^{n-r} \binom{n}{r} r!.$$

This means that

$$D_n = \sum_{r=0}^{n} (-1)^{n-r} \frac{n!}{r!(n-r)!} r! = \sum_{r=0}^{n} (-1)^{n-r} \frac{n!}{(n-r)!}$$

and so

$$p_n = \frac{D_n}{n!} = \sum_{r=0}^{n} (-1)^{n-r} \frac{1}{(n-r)!} = \sum_{s=0}^{n} (-1)^s \frac{1}{s!}$$

Table 5.2. The average number of fixed points of permutations.

n	$E(n)$
3	1.001 46...
4	1.000 48...
5	1.002 21...
10	0.997 61...
20	0.995 22...
50	1.001 97...
100	1.005 63...
1000	1.003 36...

$$p_n = 1 - \frac{1}{1!} + \frac{1}{2!} - \frac{1}{3!} + \cdots + (-1)^n \frac{1}{n!}$$

And so we have a nice result proved in three nice ways, but where is the surprise? We can reveal the first of three by looking at the average (or expected) number of correct allocations of the n objects.

The Expected Number of Fixed Points

Suppose that we perform the experiment of matching the initial arrangement $\{1, 2, 3, \ldots, n\}$ with a random permutation of itself a large number of times, and on each occasion note the number of fixed points. Each time this number of fixed points will be one of $\{0, 1, 2, \ldots, n\}$ and we can calculate the average number of them, $E(n)$. It might reasonably be thought that, as n increases, this average number increases – but consider table 5.2.

The table was constructed for each n by finding the average number of fixed points over 1000 random permutations and then averaging this number over 100 repetitions of the process: that average number of fixed points is clinging very tightly to the number 1, independently of the size of n. It could be, of course, that the program which was written to generate table 5.2 is in error, but in fact it isn't: the theoretical average number of fixed values turns out to be precisely 1 and is independent of n. This we prove below.

The expression $E(n)$ is the standard notation for the average, or expected value, and in particular of the number of fixed values of permutations of $\{1, 2, 3, \ldots, n\}$ and is defined by the standard expression

$$E(n) = \sum_{r=0}^{n} r q_r,$$

where q_r is the probability of there being precisely r fixed values, which in our case is given by

$$q_r = \frac{\binom{n}{r} D_{n-r}}{n!}.$$

The argument is that of the previous section, with the r fixed values chosen in any of $\binom{n}{r}$ ways, leaving the remaining $(n - r)$ numbers to be deranged. Our average value is, then,

$$E(n) = \sum_{r=0}^{n} r \frac{\binom{n}{r} D_{n-r}}{n!}.$$

To evaluate the expression, we will change variable by writing $s = n - r$ to get

$$E(n) = \sum_{s=0}^{n} (n - s) \frac{\binom{n}{n-s} D_s}{n!}$$

$$= \sum_{s=0}^{n} (n - s) \frac{\binom{n}{s} D_s}{n!}$$

$$= \sum_{s=0}^{n-1} (n - s) \frac{\binom{n}{s} D_s}{n!},$$

where the middle equality uses the symmetry of the binomial coefficients

$$\binom{n}{n-s} = \binom{n}{s}.$$

This means that

$$E(n) = \sum_{s=0}^{n-1} (n - s) \frac{n! D_s}{s!(n - s)! n!} = \sum_{s=0}^{n-1} \frac{D_s}{s!(n - s - 1)!}.$$

Multiplying both sides by $(n - 1)!$ results in

$$(n - 1)! E(n) = \sum_{s=0}^{n-1} \frac{(n - 1)! D_s}{s!(n - s - 1)!} = \sum_{s=0}^{n-1} \binom{n - 1}{s} D_s$$

and this expression is just equation (1) on page 53 with n replaced by $n - 1$. This means that $(n - 1)! E(n) = (n - 1)!$ and so $E(n) = 1$, independent of n.

Asymptotic Behaviour

We have, then, an attribute, $E(n)$, which is independent of the number of objects n and, if we look a little more closely at our earlier calculations, we can readily see that

$$p_n = 1 - \frac{1}{1!} + \frac{1}{2!} - \frac{1}{3!} + \cdots + (-1)^n \frac{1}{n!}$$

is also practically independent of n.

Put more positively, $1 - p_n$ is the probability of at least one match with n objects and simple computer calculations show that

$$1 - p_{13} \simeq 1 - p_{52} = 0.632\,121\ldots;$$

the two match to six decimal places, so it doesn't really matter whether we play the original Montmort version of the game or the version considered by Euler.

Figure 5.1 shows the plot of $1 - p_n$ against n as a continuous function of n with that rapid convergence very evident. Put succinctly, there is about a 63% chance of at least one match, virtually independent of n, which is perhaps higher than one might imagine.

Finally, we will look a little more closely at the series which gives p_n.

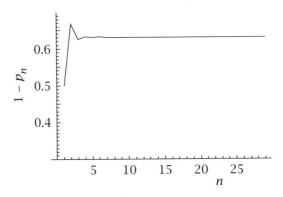

Figure 5.1. Asymptotic behaviour.

An Appearance of e

That expression

$$p_n = 1 - \frac{1}{1!} + \frac{1}{2!} - \frac{1}{3!} + \cdots + (-1)^n \frac{1}{n!}$$

has a familiar look to it and if we examine the first n terms of the Taylor expansion of

$$e^x = 1 + \frac{x}{1!} + \frac{x^2}{2!} + \frac{x^3}{3!} + \cdots$$

we can see why. The expansion is valid for all x and, in particular, for $x = -1$, and at this value the identity becomes

$$e^{-1} = 1 - \frac{1}{1!} + \frac{1}{2!} - \frac{1}{3!} + \cdots.$$

Of course, this is an infinite series and p_n has only a finite number of terms but it provides a hint that e does appear in formulae for D_n, and perhaps the nicest example of its type is

$$D_n = \left\lfloor \frac{n!}{e} + m \right\rfloor$$

and so

$$p_n = \frac{1}{n!} \left\lfloor \frac{n!}{e} + m \right\rfloor,$$

where m is any number such that $\frac{1}{3} \leqslant m \leqslant \frac{1}{2}$.

(Here, the $\lfloor \cdot \rfloor$ is the floor function defined by $\lfloor x \rfloor$ = the greatest integer less than or equal to x.)

To see this, write

$$\frac{1}{e} = \left(1 - \frac{1}{1!} + \frac{1}{2!} - \frac{1}{3!} + \cdots + (-1)^n \frac{1}{n!}\right)$$
$$+ (-1)^{n+1} \frac{1}{(n+1)!} + (-1)^{n+2} \frac{1}{(n+2)!} + \cdots ,$$

which means that

$$D_n = n!\left(1 - \frac{1}{1!} + \frac{1}{2!} - \frac{1}{3!} + \cdots + (-1)^n \frac{1}{n!}\right)$$
$$= n!\left(\frac{1}{e} - \left\{(-1)^{n+1} \frac{1}{(n+1)!} + (-1)^{n+2} \frac{1}{(n+2)!} + \cdots\right\}\right)$$
$$= \frac{n!}{e} - n!\left\{ - (-1)^n \frac{1}{(n+1)!} + (-1)^n \frac{1}{(n+2)!} - \cdots\right\}$$
$$= \frac{n!}{e} + (-1)^n \left(\frac{1}{n+1} - \frac{1}{(n+1)(n+2)} + \cdots\right)$$

and pairing the terms after the first makes clear that

$$\left|D_n - \frac{n!}{e}\right| < \frac{1}{n+1}.$$

Now, if n is even, the above expression for D_n shows that $D_n > n!/e$ and so

$$D_n = \left\lfloor \frac{n!}{e} + m \right\rfloor$$

provided that

$$\frac{1}{n+1} \leqslant m \leqslant 1$$

and since $n \geqslant 2$ we require $\frac{1}{3} \leqslant m \leqslant 1$.

If n is odd, $D_n < n!/e$ and now

$$D_n = \left\lfloor \frac{n!}{e} + m \right\rfloor$$

provided that $0 \leqslant m + 1/(n+1) \leqslant 1$ and this means that $0 \leqslant m \leqslant \frac{1}{2}$.

Take these two results together and we have the result. Of course, it can be convincingly argued that $m = 0$ provides the nicest expression

$$D_n = \left\lfloor \frac{n!}{e} \right\rfloor$$

and so

$$p_n = \frac{1}{n!} \left\lfloor \frac{n!}{e} \right\rfloor .$$

It can also be shown that

$$\frac{1}{(1-x)e^x} = \sum_{n=0}^{\infty} D_n \frac{x^n}{n!} = \sum_{n=0}^{\infty} p_n x^n$$

is a generating function for the p_n, but that is another story!

A Generalization

We have seen that derangements are permutations without a fixed point. The obvious generalization of this is to allow a specific number of fixed points, in which case we approach the general form of rencontres numbers $D_n(k)$, the number of permutations of n objects which have precisely k fixed points, $0 \leqslant k \leqslant n$. Since the work involved in finding an expression for $D_n(k)$ is now trivial, we may as well do just that. If we have k fixed points, which can be chosen in $\binom{n}{k}$ ways, we must have a derangement of the remaining $n - k$ numbers and this can be achieved in

$$\left\lfloor \frac{(n-k)!}{e} \right\rfloor$$

ways. Thus

$$D_n(k) = \binom{n}{k} \left\lfloor \frac{(n-k)!}{e} \right\rfloor .$$

And these can be conveniently displayed as a triangular array,

```
1
0     1
1     0     1
2     3     0     1
9     8     6     0     1
44    45    20    10    0     1
265   264   135   40    15    0     1
1854  1855  924   315   70    21    0     1
⋮     ⋮     ⋮     ⋮     ⋮     ⋮     ⋮     ⋮     ⋱
```

with the numbers in the leftmost vertical column the number of derangements, of course.

This means that the probability that a permutation has exactly k fixed points is

$$p_n(k) = \frac{D_n(k)}{n!} = \frac{1}{n!}\binom{n}{k}\left\lfloor \frac{(n-k)!}{e} \right\rfloor$$

$$= \frac{1}{k!}\frac{1}{(n-k)!}\left\lfloor \frac{(n-k)!}{e} \right\rfloor \xrightarrow[n\to\infty]{} \frac{e^{-1}}{k!}.$$

Summing this last expression from $k = 0$ to ∞ gives the answer 1, essential for a probability distribution. And this has connections with moments of distributions, which have connections with Bell Numbers, and these with partitions - none of which we will enter into here!

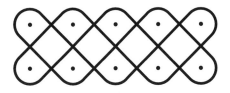

Chapter 6

CONWAY'S CHEQUERBOARD ARMY

> Games are among the most interesting creations of the human mind, and the analysis of their structure is full of adventure and surprises.
>
> James R. Newman

John Horton Conway is very hard to encapsulate. He is universally acknowledged as a world-class mathematician, a claim strongly substantiated by his occupation of the John von Neumann Chair of Mathematics at Princeton University. His vast ability and remarkable originality have caused him to contribute significantly to group theory, knot theory, number theory, coding theory and game theory (among other things); he is also the inventor of surreal numbers, which seem to be the ultimate extension of the number system and, most famous of all in popular mathematics, he invented the cellular automata game of Life. In chapter 14 we will look at him putting fractions to mysterious use, but here we will be concerned with another cellular game, typically simple, and typically deep.

The Problem

Imagine an infinite, two-dimensional chequerboard divided in half by an infinite barrier, as in figure 6.1. Above the barrier the

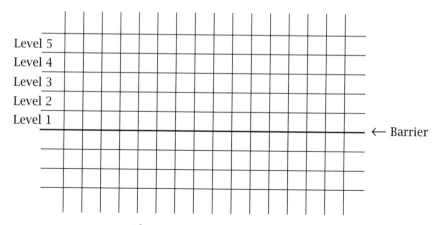

Figure 6.1. The playing area.

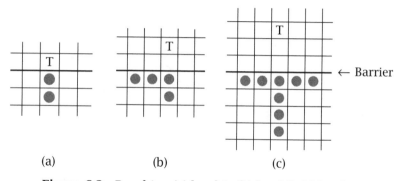

Figure 6.2. Reaching (a) level 1, (b) level 2, (c) level 3.

horizontal levels are numbered as shown. Chequers are placed on the squares below the barrier and can move horizontally or vertically below it or above it by jumping over and removing an adjacent piece. The puzzle Conway associated with this simple situation is to find starting configurations entirely below the barrier which will allow a single chequer to reach a particular target level above the barrier. It's very instructive to experiment with the pieces and, having done so, figure 6.2 shows the minimal configurations required to reach levels 1 to 3; in each case the target square T is reached by a single chequer.

The minimal number of chequers needed to reach levels 1, 2 and 3 is then 2, 4 and 8, respectively. The answer for level 4 is more complicated and, in what might be thought of as our first

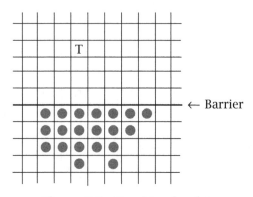

Figure 6.3. Reaching level 4.

Table 6.1. The level/chequer-count comparison.

Level	Minimum no. of chequers to reach level
1	2
2	4
3	8
4	20
5	There isn't one

surprise, figure 6.3 discloses that it is not 16 but a full 20 pieces that are needed to reach the target square T.

The second surprise, and the one which will occupy us for the rest of the chapter, is that level 5 is impossible to reach, no matter how many chequers are placed in whatever configuration below the barrier.

Table 6.1 summarizes the situation. The result is indeed surprising, but then so is Conway's ingenious method of proof, which, apart from anything else, brings in the Golden Ratio.

The Solution

To start with, fix any target square T on level 5 and, relative to it, associate with every square a nonnegative integer power of the variable x, that power being the 'chequerboard distance'

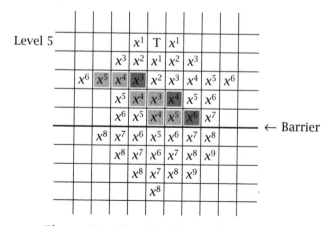

Figure 6.4. The labelling of the squares.

or 'taxicab distance' of the square from T. Such a distance is measured as the number of squares, measured horizontally and vertically from T, which gives rise to figure 6.4.

With this notation in place, every arrangement of chequer pieces, whether the initial configuration or the configuration at some later stage, can be represented by the polynomial formed by adding each of these powers of x together, for example, the starting positions to reach levels 1 to 4 might be represented by the polynomials $x^5 + x^6$, $x^5 + 2x^6 + x^7$, $x^5 + 3x^6 + 3x^7 + x^8$ and $x^5 + 3x^6 + 5x^7 + 6x^8 + 4x^9 + x^{10}$, respectively.

We now look at the effect of a move on the representing polynomial by realizing that, for this purpose, the choice of moves reduces to just three essentially different possibilities, which are characterized by the shaded cells in figure 6.4, where counters in the light grey squares are replaced by the counter in the dark grey square in each case. The general forms of these are

$$x^{n+2} + x^{n+1} \quad \text{is replaced by } x^n,$$
$$x^n + x^{n-1} \quad \text{is replaced by } x^n,$$
$$x^n + x^{n+1} \quad \text{is replaced by } x^{n+2}.$$

Any starting configuration will define a polynomial and, with every move that is made, that polynomial will change according to one of the three possibilities detailed above. The variable x

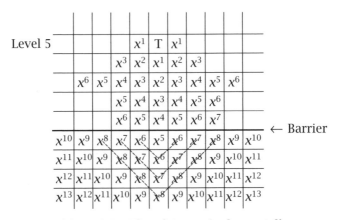

Figure 6.5. The ultimate 'polynomial'.

is arbitrary and we are free to replace it with any value we wish and will look to do so by choosing a value (greater than 0) which will cause the numeric value of the polynomial to decrease in the second and third cases and remain unchanged in the first (this last is for later algebraic convenience) when this number is substituted into it. Since $x > 0$, evidently $x^n + x^{n-1} > x^n$. If $x^n + x^{n+1} > x^{n+2}$, we require that $1 + x > x^2$ and this means that $0 < x < \frac{1}{2}(\sqrt{5} + 1) = \varphi$, which brings about the promised appearance of the Golden Ratio.

To cause the first move to leave the value of the polynomial unchanged we require that $x^{n+1} + x^{n+2} = x^n$, which means $x + x^2 = 1$ and $x = \frac{1}{2}(\sqrt{5} - 1) = 1/\varphi$, and the Golden Ratio appears once more.

So, if we make $x = 1/\varphi$ $(< \varphi)$, we are assured that the requirements are satisfied and further that, for this value of x, $x + x^2 = 1$.

Whatever our starting configuration below the dividing line, there will be a finite number of squares occupied. This means that any starting position evaluated at $x = 1/\varphi$ would be less than that of the 'infinite' polynomial generated by the occupation of every one of the infinite number of squares. We can find an expression for this by adding the terms in 'vertical darts', as illustrated in figure 6.5.

Adding terms in this way results in the expression

$$P = x^5 + 3x^6 + 5x^7 + 7x^8 + \cdots$$
$$= x^5(1 + 3x + 5x^2 + 7x^3 + \cdots).$$

The series in the brackets is a standard one (sometimes known as an arithmetic–geometric series) and is summed in the same way as a standard geometric series

$$S = 1 + 3x + 5x^2 + 7x^3 + \cdots,$$
$$\therefore \quad xS = x + 3x^2 + 5x^3 + 7x^4 + \cdots,$$
$$\therefore \quad S - xS = (1 - x)S = 1 + 2x + 2x^2 + 2x^3 + \cdots$$
$$= 1 + 2(x + x^2 + x^3 + \cdots)$$
$$= 1 + \frac{2x}{1 - x} = \frac{1 + x}{1 - x},$$
$$\therefore \quad S = \frac{1 + x}{(1 - x)^2}.$$

Multiplying by the x^5 term gives the final expression as

$$P = \frac{x^5(1 + x)}{(1 - x)^2}.$$

Since our chosen value for x satisfies $x + x^2 = x(1 + x) = 1$, it must be that $1 + x = 1/x$ and also $1 - x = x^2$. Therefore,

$$P = \frac{x^5(1/x)}{(x^2)^2} = \frac{x^5}{x^5} = 1.$$

This means that the value of any starting position must be strictly less than 1 and since each move reduces or maintains the value of the position, the value of a position can never reach 1. It is impossible, therefore, to reach level 5.

The proof can be seen to fail with the lower levels. For example, with level 4 we finish with the product

$$x^4S = x^4 \times \frac{1}{x^5} = \frac{1}{x} > 1,$$

leaving room for a reduction of the position to exactly 1.

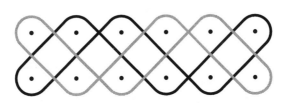

Chapter 7

THE TOSS OF A NEEDLE

Philosophy is a game with objectives and no rules.
Mathematics is a game with rules and no objectives.

Ian Ellis

The *Society for the Diffusion of Useful Knowledge*, founded (mainly by Lord Brougham) in 1828, had the object of publishing information for people who were unable to obtain formal teaching, or who preferred self-education. The celebrated English mathematician and logician Augustus De Morgan was a gifted educator who contributed no less than 712 articles to one of the society's publications, the *Penny Cyclopaedia*: one of them (published in 1838 and titled *Induction*) detailed (possibly for the first time) a rigorous basis for mathematical induction.

It would appear that De Morgan was contacted by more than his fair share of people whom we might now call mathematical cranks or, to use his own word, *paradoxers*, defined by him in the following way:

> A great many individuals, ever since the rise of the mathematical method, have, each for himself, attacked its direct and indirect consequences. I shall call each of these persons a paradoxer, and his system a paradox. I use the word in the

old sense: a paradox is something which is apart from general opinion, either in subject matter, method, or conclusion.

His unwelcome exposure to

> squarers of the circle, trisectors of the angle, duplicators of the cube, constructors of perpetual motion, subverters of gravitation, stagnators of the earth, builders of the universe...

inspired the (posthumously published) *Budget of Paradoxes*, a revised and extended collection of letters to another significant publication, the Athenæum journal.

The *Budget* is an eclectic collection of comments, opinions and reviews of 'paradoxical' books and articles which De Morgan had accumulated in his own considerable library, partly by purchase at bookstands, partly from books sent to him for review or by the authors themselves. It seems that one James Smith, a successful Liverpool merchant working at the Mersey Dock Board, was the most persistent cause of such aggravation:

> Mr. Smith continues to write me long letters, to which he hints that I am to answer. In his last of 31 closely written sides of note paper...

Mr Smith's conviction was that $\pi = 3\frac{1}{8}$ (which he seemed to 'prove' by assuming the result and showing that all other possible values then led to a contradiction). The reader may enjoy delving a little deeper into the world of mathematical cranks by reading Woody Dudley's delightful book of that name.

It is small wonder then that De Morgan picked on probability as a rich seam for the paradoxers to mine, but he recognized it as a seam which contained more than fool's gold. Again, from the *Budget* we read

> The paradoxes of what is called chance, or hazard, might themselves make a small volume. All the world understands that there is a long run, a general average; but a great part of the world is surprised that this general average should be computed and predicted. There are many remarkable cases of verification; and one of them relates to the quadrature of

the circle.... I now come to the way in which such consid-
erations have led to a mode in which mere pitch-and-toss
has given a more accurate approach to the quadrature of
the circle than has been reached by some of my paradoxers. The method is as follows: Suppose a planked floor of
the usual kind, with thin visible seams between the planks.
Let there be a thin straight rod, or wire, not so long as the
breadth of the plank. This rod, being tossed up at hazard,
will either fall quite clear of the seams, or will lay across
one seam. Now Buffon, and after him Laplace, proved the
following: That in the long run the fraction of the whole
number of trials in which a seam is intersected will be the
fraction which twice the length of the rod is of the circumference of the circle having the breadth of a plank for its
diameter. In 1855 Mr. Ambrose Smith, of Aberdeen, made
3,204 trials with a rod three-fifths of the distance between
the planks: there were 1,213 clear intersections, and 11 contacts on which it was difficult to decide. Divide these contacts equally, and we have 1,218 1/2 to 3,204 for the ratio
of 6 to 5Pi, presuming that the greatness of the number of
trials gives something near to the final average, or result
in the long run: this gives Pi=3.1553. If all the 11 contacts
had been treated as intersections, the result would have
been Pi=3.1412, exceedingly near. A pupil of mine made 600
trials with a rod of the length between the seams, and got
Pi=3.137. This method will hardly be believed until it has
been repeated so often that 'there never could have been
any doubt about it.'

We will look into this peculiar phenomenon, but first we will
mention some related games.

Fairground Games

The study of 'geometric probability', where probabilities are
determined by comparison of measurements, seems to have
had its birth in 1777 (as did the greatest of all mathematicians,
Gauss) in the paper, 'Sur le jeu de franc-carreau', published by
Georges Louis Leclerc, Comte de Buffon. The game of throwing

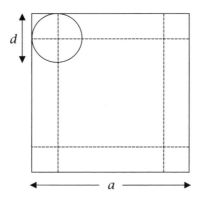

Figure 7.1. The coin on the square.

a small coin ('*un ecu*') onto a square grid was a popular pas-
time and the question of a fair fee to play the game naturally
arose; put another way, what is the probability that the coin
lands wholly in a square tile ('*à franc-carreau*')?

Buffon correctly argued that the coin would land entirely
within a square tile whenever the centre of the coin landed within
a smaller square, whose side was equal to the side of a grid
square less the diameter of the coin, as we see in figure 7.1.

If the grid square is of side a and the coin has diameter d
(which we will suppose is not greater than $\frac{1}{2}a$), this means that,
if we write this probability as p, we have

$$p = \frac{(a-d)^2}{a^2} = \left(1 - \frac{d}{a}\right)^2,$$

where $d/a \leqslant \frac{1}{2}$.

For the game to be fair, the expected value of the game must
be 0 and so, if it costs 1 unit to play and we are given w units if
we win,

$$p \times w + (1-p) \times (-1) = 0,$$

which gives

$$w = \frac{1-p}{p} = \frac{1}{p} - 1 = \frac{1}{(1-d/a)^2} - 1.$$

A plot of w against d/a is given in figure 7.2.

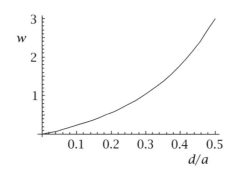

Figure 7.2. Winning behaviour.

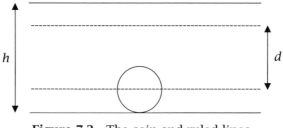

Figure 7.3. The coin and ruled lines.

To entice the player to double their money, a simple calculation shows that $d/a = 1 - 1/\sqrt{2}$, or a little less than this if we are to make a profit!

Moving from a square grid to sets of parallel lines makes the calculation even easier. If the lines are a constant distance h apart and the disc has a diameter d, it is clear from figure 7.3 that the disc will land within a pair of lines if its centre lies in a band of width $h - d$ and so the probability that this happens is

$$\frac{h - d}{h} = 1 - \frac{d}{h}.$$

Our fair game would now force

$$w = \frac{1}{p} - 1 = \frac{1}{(h-d)/h} - 1$$

$$= \frac{h}{h-d} - 1 = \frac{d}{h-d} = \frac{d/h}{1-d/h},$$

where $d < h$.

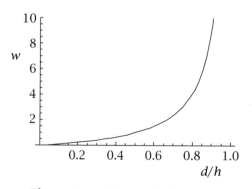

Figure 7.4. Winning behaviour.

A plot of w against d/h is given in figure 7.4.

Another simple calculation shows that, to double the stake, $d/h = 1/2$.

So far, these are geometric probabilities calculated in a reasonable manner to give reasonable answers. Now we move to the already heralded, seemingly simpler, but far more intriguing, case.

Buffon's (Short) Needle

Buffon raised the question of throwing not a circular object, but an object of a different shape, such as a square, or a 'baguette' (a rod or stick), or, as he points out, 'On peut jouer ce jeu sur un damier avec une aiguille à coudre ou une épingle sans tête.' ('You can play this game on a chequerboard with a sewing-needle or a pin without a head.') It is said that he threw a classic French baguette over his shoulder onto a boarded floor to demonstrate a version of the idea. We come, then, to the phenomenon now universally known as Buffon's Needle: if we throw a needle of length l on a board ruled with parallel lines, distance d ($\geq l$) apart, what is the probability that the needle crosses one of the lines?

In the eighteenth and nineteenth centuries such experiments were common, with probability considered as something of an experimental science. We have seen De Morgan detail the efforts of Mr Ambrose Smith of Aberdeen; this and the efforts of De Morgan himself are included in a table in the 1960 article, Geometric

Table 7.1. The number of repetitions is R, the number of crossings is C, and the estimated value of π.

Name	Date	l/d	R	C	$\sim \pi$
Wolf	1850	0.8	5000	2532	3.159 6
Smith	1855	0.6	3204	1218.5	3.155 3
De Morgan	1860	1.0	600	382.5	3.137
Fox	1864	0.75	1030	489	3.159 5
Lazzarini	1901	0.8$\dot{3}$	3408	1808	3.141 592 9
Reina	1925	0.5419	2520	869	3.179 5

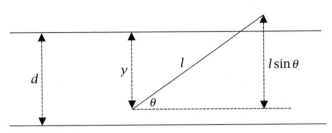

Figure 7.5. The needle crossing a line.

probability and the number π (*Scripta Mathematica* 25(3):183–95) by N. T. Gridgeman. This is reproduced in table 7.1, with the relative length of the needle and gap equal to l/d.

With all of this experimental data, it is time to look into the mathematics of all of this.

Figure 7.5 shows the needle crossing one of the horizontal lines at an angle θ to the positive x-direction. If we define y to be the distance of the lower end of the needle from the line which has been crossed, it must be that $0 \leqslant y \leqslant d$ and also $0 \leqslant \theta \leqslant \pi$. The vertical distance of the lower end to the upper end of the needle is $l \sin \theta$ and for the needle to cross the line it must be that $l \sin \theta > y$. Figure 7.6 shows a plot of the rectangular 'phase space' for the experiment, together with the curve $y = l \sin \theta$: crossings are achieved at all points underneath and on the curve.

To calculate the probability of a crossing we need to calculate the fraction

$$\frac{\text{Area below the curve}}{\text{Area of the rectangle}} = \frac{\int_0^\pi l \sin \theta \, d\theta}{\pi d} = \frac{[-l \cos \theta]_0^\pi}{\pi d} = \frac{2l}{\pi d}$$

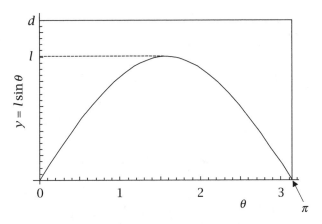

Figure 7.6. The experiment's phase space.

and so arrive at the remarkable fact that

$$\frac{2l}{\pi d} \approx \frac{C}{R}$$

and hence at an experimental method of approximating π.

If we revert to the empirical, for a given length of needle and distance between parallel lines, we can perform the experiment repeatedly in the manner of our Victorian forbears (or get a computer random number generator to do the work for us) to compute the value C/R.

In fact, 'throwing the needle' 10 000 times with $l = 1$ and $d = 2$ led to the result

$$\frac{C}{R} = 0.318\,15\ldots,$$

which, of course, means that $\pi \approx 3.143\,17\ldots$.

Buffon's (Long) Needle

The condition that $l \leqslant d$ ensures that $l\sin\theta \leqslant d$ and therefore that the curve lies within the rectangle in figure 7.6. If we wish to conduct the experiment with $l > d$, $l\sin\theta$ may well be greater than d and we will need to take into account the overlap of the curve and the rectangle, as shown in figure 7.7, and compute the area under the truncated curve.

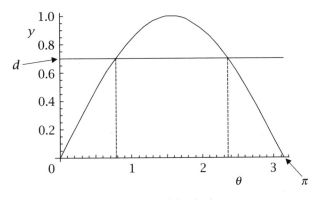

Figure 7.7. The modified phase space.

The intersections are where $l \sin \theta = d$, or $\theta = \sin^{-1}(d/l)$, and $\pi - \sin^{-1}(d/l)$. The area we want is then

$$2 \int_0^{\sin^{-1}(d/l)} l \sin \theta \, d\theta + \left\{ \left(\pi - \sin^{-1}\left(\frac{d}{l}\right) \right) - \sin^{-1}\left(\frac{d}{l}\right) \right\} d$$

$$= 2[-l \cos \theta]_0^{\sin^{-1}(d/l)} + \left\{ \pi - 2 \sin^{-1}\left(\frac{d}{l}\right) \right\} d$$

$$= 2l\left(1 - \cos\left(\sin^{-1}\left(\frac{d}{l}\right) \right) \right) + \left\{ \pi - 2 \sin^{-1}\left(\frac{d}{l}\right) \right\} d$$

$$= 2l\left(1 - \sqrt{1 - \left(\frac{d}{l}\right)^2} \right) + \left\{ \pi - 2 \sin^{-1}\left(\frac{d}{l}\right) \right\} d,$$

where the $\cos(\sin^{-1}(d/l))$ is transformed to the more convenient $\sqrt{1 - (d/l)^2}$ by use of the standard mechanism that if $\theta = \sin^{-1}(d/l)$, $\sin \theta = d/l$ and so the triangle shown in figure 7.8 exists and the third side is found by using Pythagoras's Theorem, which makes $\cos \theta = \sqrt{1 - (d/l)^2}$.

And all of this makes the probability of a crossing at least one line the rather more impressive expression

$$\frac{1}{\pi d}\left\{ 2l\left(1 - \sqrt{1 - \left(\frac{d}{l}\right)^2} \right) + \left(\pi - 2 \sin^{-1}\left(\frac{d}{l}\right) \right) d \right\}$$

$$= \frac{1}{\pi}\left\{ \frac{2l}{d}\left(1 - \sqrt{1 - \left(\frac{d}{l}\right)^2} \right) + \pi - 2 \sin^{-1}\left(\frac{d}{l}\right) \right\}.$$

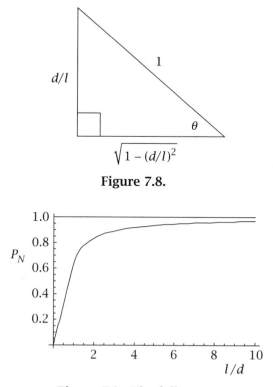

Figure 7.8.

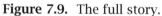

Figure 7.9. The full story.

To summarize, the probability, P_N, of the needle crossing at least one line is given by

$$P_N = \begin{cases} \dfrac{2l}{\pi d} & \text{for } l \leqslant d, \\[2ex] \dfrac{1}{\pi}\left\{\dfrac{2l}{d}\left(1 - \sqrt{1 - \left(\dfrac{d}{l}\right)^2}\right) + \pi - 2\sin^{-1}\left(\dfrac{d}{l}\right)\right\} & \text{for } l \geqslant d. \end{cases}$$

Notice that, not unreasonably, the two formulae agree at $l = d$. Figure 7.9 is a plot of this combined probability function against l/d.

The Lazzarini Entry

The fifth entry of table 7.1 stands out. The final column would have us believe that π has been estimated to an accuracy of six

Table 7.2. Lazzarini's data.

Number of repetitions R	Number of crossings C
100	53
200	107
1000	524
2000	1060
3000	1591
3408	1808
4000	2122

decimal places by the method, far in excess of the accuracy of the other entries: was it luck or deception?

In 1901 the Italian mathematician Mario Lazzarini published the result under the rather wordy title, 'Un applicazione del calcolo della probabilità alla ricerca sperimentale di un valor approssimato di π', in the journal *Periodico di Matematica* 4:140–43; four pages of fame which has led to many more pages of suspicion and of outright rebuttal. Gridgeman's article provided compelling reasons to doubt, the excellent 1965 book *Puzzles and Paradoxes* by Tim O'Beirne built on that doubt and Lee Badger's analysis in 'Lazzarini's lucky approximation of π' (*Mathematics Magazine* 67(2), April 1994) pretty much signed the intellectual death warrant.

We will not attempt to discuss the matter at any length here, but a few details from these studies are hard to ignore. In fact, Lazzarini reported the data as part of a table of results of a number of such experiments, shown as table 7.2. We must conclude that he was a patient man.

Of course, it is that penultimate entry which stands out, initially because of the curious 3408 repetitions (of a needle with $l = 2.5$ cm tossed across parallel lines with $d = 3$ cm). What also stands out is that 3.141 592 9 is the seven-decimal-place approximation to the second-best-known rational approximation of π of $\frac{355}{113}$ (known in the fifth century to Tsu Chung-chih).

Perhaps it was deception. If we follow Badger's and O'Beirne's reverse engineering, since $2l/(\pi d) \approx R/C$ it is the case that

Table 7.3. Empirical compared with theoretical data.

Number of repetitions R	Number of crossings C	Expected number of crossings C
100	53	53.05
200	107	106.10
1000	524	530.52
2000	1060	1061.03
3000	1591	1591.55
3408	1808	1808
4000	2122	2122.07

$2Cl/dR \approx \pi$ and if we use our rational estimate for π we have that

$$\frac{2Cl}{dR} \approx \frac{355}{113} = \frac{5 \times 71}{113} = \frac{5 \times 71 \times k}{113 \times k} \quad \text{for any } k.$$

A reasonable choice is $2l = 5$, which makes $l = \frac{5}{2}$ and since $d > l$ a reasonable choice for that is $d = 3$ and this makes $C/R = 213k/113k$. Provided that C and R are chosen to make their ratio $\frac{213}{113}$, the result will be achieved: with $k = 16$ we have Lazzarini's figures.

Or it may have been luck. With $2l/\pi d \approx R/C$ and $l/d = \frac{5}{6}$ we have that $5/3\pi \approx R/C$ and so $\pi \approx 5C/3R$. O'Beirne points out that one trial earlier than the given final repetition of $R = 3408$ would have $R = 3407$ and $C = 1807$ or $C = 1808$, which would make the estimate $\pi \approx 3.142\ldots$ and $\pi \approx 3.140\ldots$, respectively, each out in the third decimal place. In turn, Badger points out that had there been $C = 1807$ or $C = 1809$ crossings in 3408 repetitions, the estimates would be $\pi \approx 3.143\ldots$ and $\pi \approx 3.139\ldots$, respectively; the experiment does seem to have stopped on something of a cusp of luck.

Now consider the rest of the data. Again, with $l/d = \frac{5}{6}$, the probability of a crossing is $2l/\pi d = 5/3\pi$ and so, on average, the expected number of crossings is $(5/3\pi) \times R$ and if we extend table 7.2 to include these values we arrive at table 7.3.

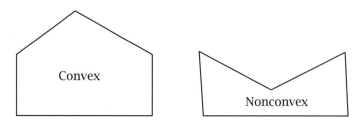

Figure 7.10. Convexity and nonconvexity.

It all looks too accurate and some simple statistical tests quantify that suspicion; the chances of this happening are less than 3×10^{-5}. And this is our, but by no means Badger's, final word on the topic.

A Generalization and a Final Surprise

Buffon mentioned throwing a square object: in fact, we can formulate a surprising result for any convex, polygonal lamina. First, a polygonal lamina is *convex* if it contains all line segments connecting all pairs of points on it. For example, figure 7.10 shows a convex and a nonconvex pentagon.

Note that an immediate consequence of convexity is that any straight line will intersect precisely two sides of the lamina or none at all.

Now suppose that we throw the lamina onto a set of parallel, ruled lines a constant distance d apart. Suppose also that the lamina is made up of n sides of length l_i for $i = 1, 2, \ldots, n$, where each side is less than d. Since the order in which we count the sides is irrelevant, the intersection of a ruled line with the lamina must occur with the line pair $(l_i l_j)$ of the lamina for some pair i and j, where we may assume $i < j$, and suppose that this occurs with probability $P(l_i l_j)$. This means that the probability of an intersection of the lamina with a line is $P = \sum_{i<j} P(l_i l_j)$.

If the side l_i is intersected with probability $P(l_i)$, since the lamina is convex, so must exactly one of the remaining sides, and so $P(l_i) = \sum_{j \neq i} P(l_i l_j)$. This means that

$$\sum_{i=1}^{n} P(l_i) = \sum_{i,j:i \neq j} P(l_i l_j) = 2 \sum_{i,j:i<j} P(l_i l_j) = 2P.$$

Now we use the previous result for Buffon's Needle to write $P(l_i) = 2l_i/\pi d$ and so $2P = \sum_{i=1}^{n} 2l_i/\pi d$, which makes

$$P = \frac{1}{\pi d} \sum_{i=1}^{n} l_i = \frac{1}{\pi d} \times (\text{Perimeter of lamina});$$

the probability that the lamina crosses a line is completely independent of its shape, depending only on its perimeter.

There are many more variants and generalizations of the original, novel idea of the eighteenth-century polymath Georges Louis Leclerc, Comte de Buffon: instead of parallel lines a rectangular grid, or perhaps radial lines or unequally spaced lines with a needle with a 'preferred' orientation (which is apparently useful in determining the spacing of flight lines for locating anomalies in airborne geophysical surveys). The Monte Carlo technique, of which this is the original example, is commonly used to estimate lengths of curves and areas of regions. In recent research on ants choosing nesting sites it has been suggested that the ant scouts' critical job of site selection is influenced by estimates of area based on a variant of Buffon's principle. Newton's words have resonance:

> Nature is pleased with simplicity, and affects not the pomp of superfluous causes.

But then, with so much needle tossing to do, so have those of Buffon himself:

> Never think that God's delays are God's denials. Hold on; hold fast; hold out. Patience is genius!

Chapter 8

TORRICELLI'S TRUMPET

The notion of infinity is our greatest friend; it is also the greatest enemy of our peace of mind.

James Pierpont

An Argument

One of the longest and most vitriolic intellectual disputes of all time took place between the two seventeenth-century luminaries Thomas Hobbes and John Wallis: Hobbes, the philosopher, had claimed to have 'squared the circle' and Wallis, the mathematician, had strongly and publicly refuted that claim.

This ancient problem (one of three of its kind) had been handed down by the Greeks and asked if it was possible, using straight edge and compasses only, to construct a square equal in area to the given circle: it took until 1882 until Ferdinand Lindemann proved π to be transcendental, which meant that the question was resolved in the negative. Wallis was right.

Although the 'squaring the circle' problem spawned the conflict, the battle lines extended far beyond it – and in fact to the infinite, a concept which was far from understood at the time and which brought with it all manner of technical and philosophical difficulties, and it is one particular example of infinity's capricious nature that crystallized the adversaries' opposing views of the concept: Torricelli's Trumpet (or The Archangel Gabriel's Trumpet, or Horn).

A Strange Trumpet

Bonaventura Cavalieri was a mathematician good enough to be praised by Galileo, who said of him, 'few, if any, since Archimedes, have delved as far and as deep into the science of geometry'. And it was Archimedes' *method of exhaustion* which Cavalieri developed to form his theory of *indivisibles*, that is, finding lengths, areas and volumes by slicing the object in question into infinitesimally small pieces. To this he added *Cavalieri's Principle*. This was 1629 and integral calculus was yet to be forged at the anvils of the yet unborn Newton and Leibniz; Cavalieri's ideas would help with the process.

Evangelista Torricelli (remembered as the inventor of the barometer), frequent correspondent of Cavalieri and assistant to an ageing Galileo, was also an accomplished mathematician. Most particularly, using the method of *indivisibles*, in 1645 he *rectified* the Logarithmic Spiral (that is, he was able to measure the length of the curve; a result we will use in chapter 10).

Our interest lies with some of his earlier work when, in 1643, he had made known his discovery of the strange nature of the *acute hyperbolic solid*, which we would now call the rectangular hyperboloid. It is generated by rotating the rectangular hyperbola $y = 1/x$ by 360° about the x-axis. Figure 8.1 shows the solid.

He showed that this infinite solid has a finite volume. To today's post-calculus eyes this single fact is not shocking but it does become rather more surprising when we realize that not only is its length infinite, but so is its surface area.

We will use calculus and modern-day notation to prove both results: that the volume is finite and the surface area infinite. First, we will take the trouble to demonstrate Torricelli's method of showing that the volume is finite, a result which shocked the thinkers of the day.

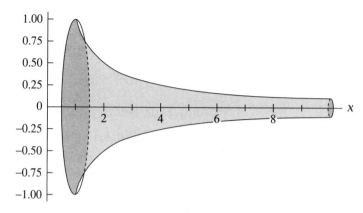

Figure 8.1. Torricelli's Trumpet.

Torricelli's Proof

We have mentioned the term *Cavalieri's Principle.* Archimedes had used the idea to find the volume of a sphere and a version of it can be stated in the following way:

> Given two solids included between parallel planes, if every plane cross-section parallel to the given planes has areas in the same ratio in both solids, then the volumes of the solids are in that ratio.

In particular, if the areas of the sections are always equal, then so are the volumes.

The principle is deceptively powerful and before we discuss Torricelli's use of it with the trumpet we will acclimatize ourselves to it by looking at a famous example of its use: computing the volume of a sphere (knowing the volume of a circular cone).

On the left of figure 8.2 is a hemisphere of radius r with a horizontal section at height h above the base of the hemisphere. On the right is a cylinder of radius r and of height r with another horizontal section cut at height h above its base. Within the cylinder is inscribed a circular cone with base the top of the cylinder and vertex at the centre of the cylinder's base. The area of the horizontal circular section within the hemisphere is $\pi(r^2 - h^2)$. Since the height of the cone at this level is h, so must be its base radius. This means that the area of the annular region within

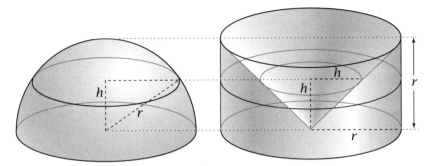

Figure 8.2. Cavalieri's Principle at work.

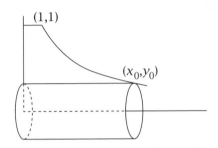

Figure 8.3. Torricelli's use of Cavalieri's Principle
which demonstrated the infinite volume.

the cylinder is $\pi r^2 - \pi h^2 = \pi(r^2 - h^2)$ also. Using Cavalieri's
Principle it must be that the volume of the hemisphere is equal
to the volume of the cylinder minus the volume of the cone. We
have, then, that the volume of the hemisphere is

$$\pi r^2 \times r - \tfrac{1}{3}\pi r^2 \times r = \tfrac{2}{3}\pi r^3$$

and the volume of the sphere is

$$\tfrac{4}{3}\pi r^3.$$

Torricelli used an extended form of the principle to show that
the volume of the infinite trumpet is itself finite. To appreciate
his proof we must imagine the trumpet to have a 'lip' at its open
end and be made up of an infinite number of concentric, horizontal cylinders. Figure 8.3 shows a particular example of such a
cylinder and let us suppose that the lip begins at the point $(1, 1)$.

(a)   (b)

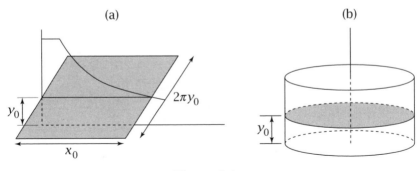

Figure 8.4.

Suppose now that the right end of a particular cylinder touches the hyperbola in the xy-plane at the point (x_0, y_0), then the cylinder has curved surface area $(2\pi y_0)x_0 = 2\pi x_0 y_0 = 2\pi$ since $y_0 = 1/x_0$. This area is, therefore, constant as the point of contact varies. Now unravel the cylinder so that it becomes a horizontal rectangle located at height y_0, as shown in figure 8.4(a). Finally, construct a vertical cylinder made up of horizontal discs each of area 2π at height y_0 from its base, as shown in figure 8.4(b). The height of this cylinder will be 1. Cavalieri's Principle is itself a limiting argument, if that is now extended to allow $x_0 \rightarrow \infty$ and so approach the base of the vertical cylinder, the volume of the trumpet is the same as the volume (2π) of that cylinder.

Of course, this argument is special. It works because $xy = 1$ and could not be easily extended to other cases, but it does work and it did shock. Torricelli himself said

> It may seem incredible that although this solid has an infinite length, nevertheless none of the cylindrical surfaces we considered has an infinite length but all of them are finite.

If we look at the problem through modern eyes, taking the finite trumpet from $x = 1$ to $x = N$ (forgetting the lip) and then allowing N to become arbitrarily large, we have that the volume of the trumpet is given by

$$\pi \int_1^N \left(\frac{1}{x}\right)^2 dx = \pi \int_1^N \frac{1}{x^2} dx = \pi \left[-\frac{1}{x}\right]_1^N = \pi\left(1 - \frac{1}{N}\right),$$

and as $N \rightarrow \infty$ the volume approaches π.

The simple calculation

$$\int_1^N \frac{1}{x}\, dx = [\ln x]_1^N = \ln N \to \infty \quad \text{as } N \to \infty$$

shows that we have a solid of infinite cross-sectional area but finite volume.

To calculate the surface area of the finite solid and so prove that it is also infinite requires more effort.

The Trumpet's Surface Area

Using the standard formula for the surface area of a volume of revolution, as described in appendix C:

$$\text{Surface area} = 2\pi \int_1^N y\sqrt{1 + \left(\frac{dy}{dx}\right)^2}\, dx.$$

Here we have $y = 1/x$ and so

$$\frac{dy}{dx} = -\frac{1}{x^2},$$

so the formula becomes

$$\text{Surface area} = 2\pi \int_1^N \frac{1}{x}\sqrt{1 + \frac{1}{x^4}}\, dx.$$

This clearly diverges since

$$\int_1^N \frac{1}{x}\sqrt{1 + \frac{1}{x^4}}\, dx > \int_1^N \frac{1}{x}\, dx$$

but it is pleasing, if a little messy, to find an exact form for the integral

$$\int_1^N \frac{1}{x}\sqrt{1 + \frac{1}{x^4}}\, dx = \int_1^N \frac{\sqrt{x^4 + 1}}{x^3}\, dx.$$

We will attack it in two stages: first using integration by parts and then substitution. So, with the indefinite integral:

$$\int \frac{\sqrt{x^4+1}}{x^3}\,dx = \int x^{-3}\sqrt{x^4+1}\,dx$$

$$= \frac{x^{-2}}{-2}\sqrt{x^4+1} - \int \frac{x^{-2}}{-2} \times \frac{1}{2} \times \frac{4x^3}{\sqrt{x^4+1}}\,dx$$

$$= -\frac{1}{2x^2}\sqrt{x^4+1} + \int \frac{x}{\sqrt{x^4+1}}\,dx.$$

Now consider the remaining integral and use the substitution $u = x^2$, in which case $du/dx = 2x$. We then have

$$\int \frac{x}{\sqrt{x^4+1}}\,dx = \frac{1}{2}\int \frac{1}{\sqrt{u^2+1}}\,du$$

$$= \tfrac{1}{2}\ln(u + \sqrt{u^2+1}) + c$$

$$= \tfrac{1}{2}\ln(x^2 + \sqrt{x^4+1}) + c.$$

Putting all this together results in the surface area of the trumpet being given by

$$\left[-\frac{1}{2x^2}\sqrt{x^4+1} + \tfrac{1}{2}\ln(x^2 + \sqrt{x^4+1}) \right]_1^N$$

$$= -\frac{1}{2N^2}\sqrt{N^4+1} + \tfrac{1}{2}\ln(N^2 + \sqrt{N^4+1}) + \tfrac{\sqrt{2}}{2} - \tfrac{1}{2}\ln(1 + \sqrt{2}).$$

As $N \to \infty$ the first term clearly approaches $-\tfrac{1}{2}$ but the log function increases without bound, which means that the surface area also increases without bound.

   (It is appropriate that an anagram of 'Evangelista Torricelli' is 'Lo! It is a clever integral'.)

The Trumpet's Centre of Mass

The confusion is complete when we consider a comment of Wallis that a

> surface, or solid, may be supposed to be so constituted, as to be Infinitely Long, but Finitely Great, (the Breadth

continually decreasing in greater proportion than the Length Increaseth,) and so as to have no Centre of Gravity. Such is Toricellio's Solidum Hyperbolicum acutum.

Using the standard calculus definition of the centre of mass $\bar{x}$ of a solid of revolution about the x-axis, we have that

$$\left(\pi \int_1^N y^2 \, dx\right)\bar{x} = \pi \int_1^N xy^2 \, dx,$$

and we have in our case

$$\pi\left(1 - \frac{1}{N}\right)\bar{x} = \pi \int_1^N x\left(\frac{1}{x}\right)^2 dx = \pi \int_1^N \frac{1}{x} \, dx$$

$$= \pi[\ln x]_1^N = \pi \ln N,$$

$$\bar{x} = \frac{\ln N}{1 - 1/N} \to \infty \quad \text{as } N \to \infty.$$

A Drinking Vessel

So, in 1643 Torricelli brought to the mathematical world a solid which has infinite surface area but finite volume. In 1658 Christiaan Huygens and René François de Sluze added to the mathematical unease of the time by reversing the conditions: their solid has finite surface area and infinite volume.

We will not consider their arguments or more modern ones to establish the fact, but the solid is generated from the cissoid (meaning 'ivy-shaped'). The canonical curve has equation $y^2 = x^3/(1 - x)$, which is shown in figure 8.5; evidently, it has a vertical asymptote at $x = 1$. It is attributed to Diocles in about 180 BC in connection with his attempt to duplicate the cube by geometrical methods and appears in Eutocius's commentaries on Archimedes' *On the Sphere and the Cylinder*, wherein *the method of exhaustion* was developed. The solid concerned is contained between the rotation of the upper half of the cissoid and the vertical asymptote about the y-axis; it forms a goblet-shaped figure, as shown in figure 8.6.

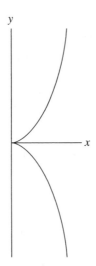

Figure 8.5.

Figure 8.6. The cissoid.

In a letter to Huygens, de Sluze mischievously described the
solid as

> a drinking glass that had small weight, but that even the
> hardiest drinker could not empty

(levi opera deducitur mensura vasculi, pondere non magni, quod
interim helluo nullus ebibat).

Torricelli's Trumpet would satisfy the more moderate drinker,
but the glass could never be wetted! Admittedly, this is fanciful

for several important reasons, but the imagery is compelling. Where is the paradox? As ever, our senses have deceived us when we have brought about the confusion which arises when we try to bring to the real world something which cannot exist within it; infinitely long things cannot be brought into reality (Euclid's parallels postulate reveals the danger in trying to do so) and wine is not infinitely thin.

Galileo's own view echoes this:

> [Paradoxes of the infinite arise] only when we attempt, with our finite minds, to discuss the infinite, assigning to it those properties which we give to the finite and limited.

But we should leave the last word to Hobbes, when he commented on the assertion that an infinite solid of finite volume exists:

> To understand this for sense, it is not required that a man should be a geometrician or a logician, but that he should be mad.

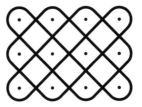

Chapter 9

NONTRANSITIVE EFFECTS

We want the surprise to be transitive like the impatient thump which unexpectedly restores the picture to the television set, or the electric shock which sets the fibrillating heart back to its proper rhythm.

Seamus Heaney

The Background

A dictionary definition of the adjective 'transitive' is 'being or relating to a relationship with the property that if the relationship holds between a first element and a second and between the second element and a third, it holds between the first and third elements.'

Initially it is easy to imagine that all meaningful relationships between all pairs of objects are transitive: 'older than', 'bigger than', etc., but we do not need to look too far to produce examples for which transitivity fails: 'son of', 'perpendicular to', etc. This chapter is primarily concerned with a relationship which is seemingly transitive, but in fact need not be: that relationship is 'better than'.

For example, suppose that A is a better tennis player than B and that B is a better tennis player than C, then it's pretty clear

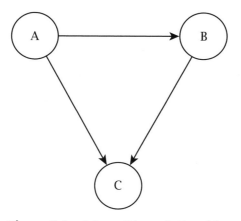

Figure 9.1. A transitive relationship.

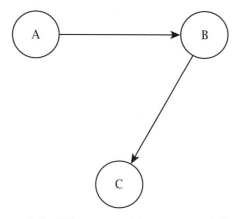

Figure 9.2. Where transitivity is not defined.

that A is a better tennis player than C; 'better than' is evidently transitive in this case.

If we represent a relationship between pairs of elements by '→' and call the elements A, B and C, it will be transitive if figure 9.1 holds.

With the two nontransitive relationships above, it is the case that the diagram simply cannot be completed and so becomes figure 9.2; A and C simply do not share the relationship of A with B and B with C.

Things become decidedly more confusing when the diagram completes to figure 9.3, where the arrows chase each other's

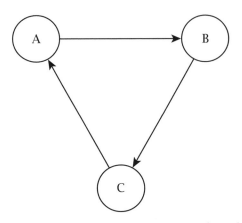

Figure 9.3. Where transitivity is confounded.

tails; this is altogether stranger. In particular, how can it be that A is better than B, B is better than C and yet C is better than A?

Do such relationships exist? The answer is decidedly *yes*. A familiar example from childhood is the game of scissors–paper–rock, where each of two players holds a hand behind his or her back. On the count of three, both players bring their hidden hand forward in one of three configurations. Two fingers in a 'V' to represent scissors, the whole hand flat and slightly curved to represent paper, and a clenched fist to represent rock. The winner is determined by the following sequence of rules: scissors cut paper, paper wraps rock and rock breaks scissors, where 'better than' has an appropriate definition in each of the three cases. There is no 'best' choice and, with A representing 'scissors', B representing 'rock' and C representing 'paper', that tail-chasing is evident.

We will continue to more devious examples; in each case, 'better than' is given the specific interpretation 'is more likely to win than'.

The Lo Shu Magic Square

The 4200-year-old Lo Shu magic square, shown in figure 9.4, provides the basis of the first example and one can be confident that the mathematicians of the time of Emperor Yu would have had no idea of this hidden property of the design. It is, of course, a

4	9	2
3	5	7
8	1	6

Figure 9.4. The Lo Shu magic square.

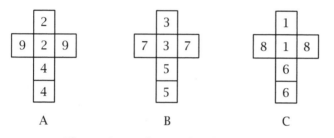

A   B   C

Figure 9.5. The Lo Shu dice nets.

3×3 square with each of the nine squares filled with one of the integers from 1 to 9, the 'magic' stems from the fact that each row, column and diagonal add up to the magic number of 15.

Now take the three rows and number three six-sided die each with two repeats of the three numbers forming each row, as shown by the nets of figure 9.5.

We can use these curiously numbered dice to play a simple game of chance with an opponent: he chooses a die, then we choose a die and we roll them (say) 100 times and see who wins the most times. Table 9.1 lists the possible outcomes with each die matched against each, and shows that A → B → C → A, each with a probability of $\frac{20}{36} = \frac{5}{9}$. We have a situation which is modelled by figure 9.3, which means that, if we allow our opponent first choice of die, we will always be in the better position.

The choice of numbers is not unique. Toy collector and consultant Tim Rowett devised a set of three nontransitive dice where no face has a number higher than 6 (the highest number on a standard six-sided die); figure 9.6 gives the nets. Again, A → B → C → A; in this case, the reader may wish to check that the probabilities of winning in each case are $\frac{25}{36}$, $\frac{21}{36}$, $\frac{21}{36}$, respectively.

Table 9.1. The Lo Shu dice compared.

	A					
B	3	3	5	5	7	7
2	A	A	A	A	A	A
2	A	A	A	A	A	A
4	B	B	A	A	A	A
4	B	B	A	A	A	A
9	B	B	B	B	B	B
9	B	B	B	B	B	B

	B					
C	2	2	4	4	9	9
1	B	B	B	B	B	B
1	B	B	B	B	B	B
6	C	C	C	C	B	B
6	C	C	C	C	B	B
8	C	C	C	C	B	B
8	C	C	C	C	B	B

	C					
A	1	1	6	6	8	8
3	A	A	C	C	C	C
3	A	A	C	C	C	C
5	A	A	C	C	C	C
5	A	A	C	C	C	C
7	A	A	A	A	C	C
7	A	A	A	A	C	C

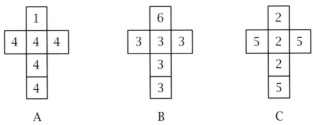

Figure 9.6. The Rowett dice nets.

Table 9.2. Possibilities when the Rowett dice are thrown twice.

A		B		C	
Total	Frequency	Total	Frequency	Total	Frequency
2	1	6	25	4	9
5	10	9	10	7	18
8	25	12	1	10	9

Table 9.3. The Rowett dice compared again.

		A		
B		2	5	8
6		B: $1 \times 25 = 25$	B: $10 \times 25 = 250$	A: $25 \times 25 = 625$
9		B: $1 \times 10 = 10$	B: $10 \times 10 = 100$	B: $25 \times 10 = 250$
12		B: $1 \times 1 = 1$	B: $10 \times 1 = 10$	B: $25 \times 1 = 25$

		B		
C		6	9	12
4		B: $25 \times 9 = 225$	B: $10 \times 9 = 90$	B: $1 \times 9 = 9$
7		C: $25 \times 18 = 450$	B: $10 \times 18 = 180$	B: $1 \times 18 = 18$
10		C: $25 \times 9 = 225$	C: $10 \times 9 = 90$	B: $1 \times 9 = 9$

		C		
A		4	7	10
2		C: $9 \times 1 = 9$	C: $18 \times 1 = 18$	C: $9 \times 1 = 9$
5		A: $9 \times 10 = 90$	C: $18 \times 10 = 180$	C: $9 \times 10 = 90$
8		A: $9 \times 25 = 225$	A: $18 \times 25 = 450$	C: $9 \times 25 = 225$

To add to the confusion, alter the game to one in which one of the dice is chosen by each player and thrown twice with the winner the person with the higher total. Table 9.2 gives the three possible totals for each die and the frequency with which each occurs.

Now, if we perform the calculations as before, we arrive at table 9.3, which shows the arrows of dominance are reversed. That is, A → C → B → A and this time with probabilities $\frac{765}{1296}$, $\frac{765}{1296}$, $\frac{671}{1296}$, respectively.

Table 9.4. Effron's dice compared.

			A			
B	1	2	3	9	10	11
0	A	A	A	A	A	A
1	X	A	A	A	A	A
7	B	B	B	A	A	A
8	B	B	B	A	A	A
8	B	B	B	A	A	A
9	B	B	B	X	A	A

			B			
C	0	1	7	8	8	9
5	C	C	B	B	B	B
5	C	C	B	B	B	B
6	C	C	B	B	B	B
6	C	C	B	B	B	B
7	C	C	X	B	B	B
7	C	C	X	B	B	B

			C			
D	5	5	6	6	7	7
3	C	C	C	C	C	C
4	C	C	C	C	C	C
4	C	C	C	C	C	C
5	X	X	C	C	C	C
11	D	D	D	D	D	D
12	D	D	D	D	D	D

			D			
A	3	4	4	5	11	12
1	D	D	D	D	D	D
2	D	D	D	D	D	D
3	X	D	D	D	D	D
9	A	A	A	A	D	D
10	A	A	A	A	D	D
11	A	A	A	A	X	D

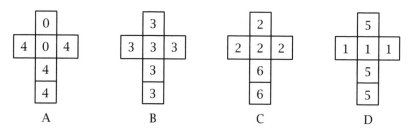

Figure 9.7. Effron's dice nets 1.

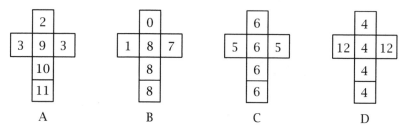

Figure 9.8. Effron's dice nets 2.

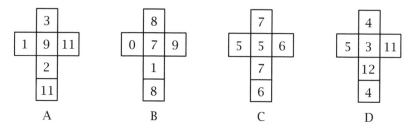

Figure 9.9. Effron's dice nets 3.

Effron's Dice

Bradley Effron, a statistician at Stanford University, extended the idea to four dice, giving the specification for three such sets, as shown in figures 9.7–9.9. In each case, A → B → C → D → A.

This is a little more subtle since the possibility of draws (re-throws) exists and we will take the trouble to compile the tables, shown as table 9.4. In each table, the event of matching numbers is represented by an X.

The ambiguity is exactly the same in all four cases and we can representatively deal with just the first, with A competing with

Table 9.5. Coin-tossing comparisons.

B chooses	A chooses	Probability of A winning
HHH	THH	$\frac{7}{8}$
HHT	THH	$\frac{3}{4}$
HTH	HHT	$\frac{2}{3}$
HTT	HHT	$\frac{2}{3}$
THH	TTH	$\frac{2}{3}$
THT	TTH	$\frac{2}{3}$
TTH	HTT	$\frac{3}{4}$
TTT	HTT	$\frac{7}{8}$

B. If we write p for the probability that A wins, we have

$$p = \frac{22}{36} + \frac{1}{36} \times p + \frac{1}{36} \times p,$$

which makes $p = \frac{11}{17}$ and so A → B with a probability of $\frac{11}{17}$ and, of course, the probability is the same for the other pairings.

Coin Tossing

The second type of nontransitive effect that we will consider involves the spin of a fair coin. Inevitably, Martin Gardner has considered it, but the author first came across the phenomenon in the Warwick University mathematics magazine *Manifold*, which has long since disappeared. Player A takes a fair coin and repeatedly spins it but before doing so asks player B to choose a sequence of three heads and tails, for example, HTH. Having done so, A chooses his own sequence. The coin is repeatedly spun until one of the two sequences appears: whosoever's sequence it is, wins. There are only eight possible choices for the triplet and B might reasonably think that somewhere among them there is a best choice, but there isn't.

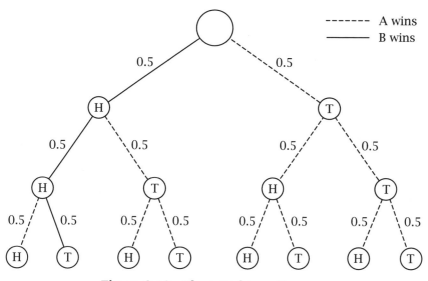

Figure 9.10. The initial tree diagram.

The left-hand column of table 9.5 shows the eight possible choices that B can make. The middle column shows the corresponding choice that A should make in each case. If he makes that choice, the right-hand column shows the probability of A winning.

Compiling the column of odds takes a little effort and makes ample use of tree diagrams. We will consider the essentially different pairings separately, dealing in detail with the first of them.

The Pairings HHH ↔ THH *and* TTT ↔ HTT

The first three tosses could be HHH, in which case B wins. Otherwise, a tail will appear among them and if this is the case, no matter how many more tails appear, A needs two heads and B still needs all three heads to win. A will assuredly get his two heads before B gets the three and it is just a mater of time before A wins. So, of the eight possibilities for the first three tosses, A will win on seven of them and so the probability of A winning is $\frac{7}{8}$.

This is easy. Now to a slightly harder case.

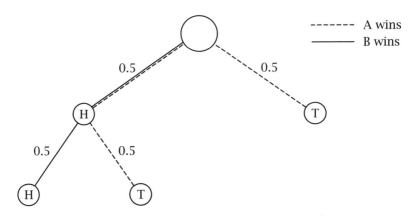

Figure 9.11. The tree diagram pruned.

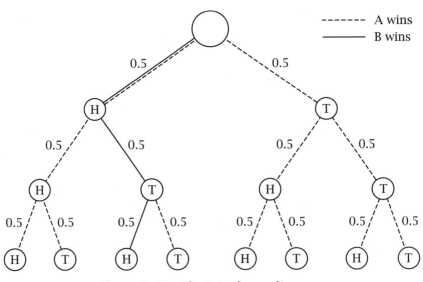

Figure 9.12. The initial tree diagram.

The Pairings HHT ↔ THH *and* TTH ↔ HTT

We will analyse the situation using a three-stage tree diagram, as shown in figure 9.10.

Figure 9.10 shows the two paths along which either A or B clearly wins, but a little thought allows us to prune the tree diagram. First, if the first toss is a T, no matter what happens subsequently, HH is needed to complete the sequence and these will

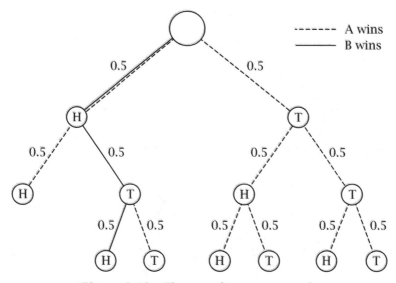

Figure 9.13. The tree diagram pruned.

only start the sequence chosen by B; A must win; this means that the whole right-hand side of the tree diagram is a win for A. Further, if the first two tosses are HT, then A must win for the same reason. Lastly, if the first two tosses are HH, then B must win. The tree diagram reduces to figure 9.11.

The probability that A wins is, then, $\frac{1}{2} + \frac{1}{2} \times \frac{1}{2} = \frac{3}{4}$.

Finally, we need to deal with the middle four cases, and this is rather more subtle since the tree diagram does not fully resolve the situation in all cases.

The Pairings HTH ↔ HHT, HTT ↔ HHT,
THH ↔ TTH *and* THT ↔ TTH

The three-stage tree diagram is shown in figure 9.12

Once again, the branches can be trimmed, but by not so much this time. If the coin comes up HH, then A will assuredly win and similarly if it comes up THH. This results in figure 9.13.

To better analyse the remaining possibilities it is useful to add in an extra level, as in figure 9.14, where p is the probability that A wins.

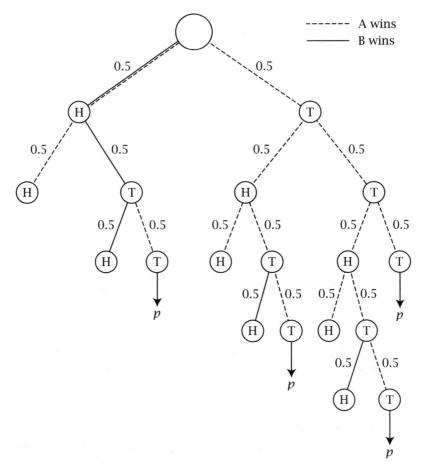

Figure 9.14. The pruned tree diagram extended.

If we work through the branches from left to right, we have the expression

$$p = (\tfrac{1}{2} \times \tfrac{1}{2}) + (\tfrac{1}{2} \times \tfrac{1}{2} \times \tfrac{1}{2} \times p) + (\tfrac{1}{2} \times \tfrac{1}{2} \times \tfrac{1}{2} + \tfrac{1}{2} \times \tfrac{1}{2} \times \tfrac{1}{2} \times \tfrac{1}{2} \times p)$$
$$+ (\tfrac{1}{2} \times \tfrac{1}{2} \times \tfrac{1}{2} \times \tfrac{1}{2}) + (\tfrac{1}{2} \times \tfrac{1}{2} \times \tfrac{1}{2} \times \tfrac{1}{2} \times \tfrac{1}{2} \times p)$$
$$+ (\tfrac{1}{2} \times \tfrac{1}{2} \times \tfrac{1}{2} \times p),$$

which makes $\tfrac{21}{32} \times p = \tfrac{7}{16}$ and so $p = \tfrac{2}{3}$.

The analysis is complete and we have established the nontransitivity.

Chapter 10

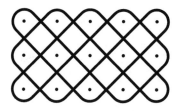

A PURSUIT PROBLEM

This book is written in mathematical language and its characters are triangles, circles and other geometrical figures, without whose help... one wanders in vain through a dark labyrinth.

Galileo Galilei

The defunct magazine *Graham DIAL* was circulated to 25 000 American engineers during the 1940s and featured a *Private Corner for Mathematicians*, edited by L. A. Graham himself and populated by problems posed by readers for other readers to solve. Akin to Martin Gardner's articles in *Scientific American*, the articles spawned two books which discussed, commented on and sometimes extended the original contributions. The first book, *Ingenious Mathematical Problems and Methods*, was published in 1959 and contains the problem we will discuss here. It is framed as a chase on the high seas, and does not seem to have sufficient information provided to be able to solve it. For its solution we will need two special curves, the nature of which we will deal with first.

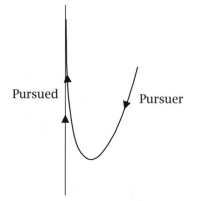

Figure 10.1. A linear pursuit curve.

A Linear Pursuit Curve

Pursuit curves were first studied in 1732 by the French scientist Pierre Bouguer, who was also the first person to measure the Earth's magnetic field. Their exact nature depends on the path of the pursued and the method of pursuit, but the common ground is that they are the paths a pursuer should take when attempting to intercept a quarry.

Suppose that we assume the pursued to be moving in a straight line and that the pursuer always steers towards his quarry's current position, continually altering course to achieve this. With this agreed, we arrive at the 'linear pursuit curve', solved by Arthur Bernhart, and which can be shown to have an equation of the form $y = cx^2 - \ln x$; it is shown in figure 10.1.

Pursuit Using the Circle of Apollonius

Alternatively, the pursuer could catch the quarry more quickly by utilizing a special plane curve: the Circle of Apollonius, named after Apollonius of Perga (ca. 262 B.C.E. to ca. 190 B.C.E.), which can be defined in the following way.

Take two distinct fixed points A and B and consider the set of all points P such that $PA{:}PB = k$ for some positive constant k. If $k = 1$, the points form the perpendicular bisector of AB, otherwise they form a circle, the Circle of Apollonius, as shown in figure 10.2.

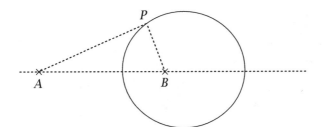

Figure 10.2. The circle of Apollonius.

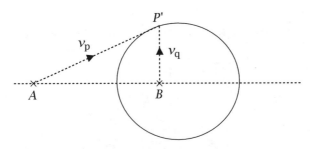

Figure 10.3. A certain capture.

Now suppose that the speed of the pursuer is v_p and that of the quarry is v_q and that at some point the pursuer is at position A and the quarry at position B. Given that the pursuer knows v_q, he should mentally construct the Circle of Apollonius as the set of points P such that $AP:PB = v_p:v_q$, as shown in figure 10.3. Given also that the pursuer knows the direction of flight, he will be able to calculate the point P' at which the pursued will cross the circle. He should head for P' and so ensure capture at that point.

The Circle of Apollonius will not be sufficient in itself for the solution to our problem, but it will have its contribution to make.

Pursuit Using the Logarithmic Spiral

A pleasing generalization of a pursuit curve results from taking several objects, each one acting as pursuer and quarry. For example, in figure 10.4 we imagine four spiders, each starting at a corner of a square and moving with equal, constant speed toward one another. In figure 10.5 lines have been drawn linking

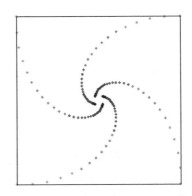

Figure 10.4. Four spiders in mutual pursuit.

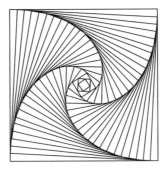

Figure 10.5. Four spiders with some links.

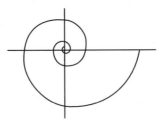

Figure 10.6. A logarithmic spiral.

some of the positions of pursuer and quarry and the diagram becomes a work of mathematical art.

These last, artistically satisfying, examples trace self-similar curves known as *logarithmic* (or equiangular) *spirals*, whose polar equation is $r = ae^{b\theta}$ for constants a and b. A typical example is shown in figure 10.6. They were first studied in 1638 by

René Descartes but are most famously associated with Jakob Bernoulli, who developed many of their startling properties. So enamoured of them was he that he asked for one to be engraved on his tombstone with the phrase 'Eadem mutata resurgo' ('I shall arise the same, though changed'); unfortunately, the stone-mason seems to have been unduly challenged by the charge and produced a somewhat crude Archimedean Spiral (whose polar form is $r = a\theta$). An essential difference between the two spirals is that successive turnings of the Archimedean Spiral have a constant separation distance (of $2\pi a$), whereas with the logarithmic spiral these distances are in geometric progression.

Logarithmic spirals abound in nature: they are the paths along which insects approach a light source and hawks approach their prey, the shape of spiral galaxies (including our own Milky Way) and also of cyclones. In Book 1 of *Principia* Newton proved that if the universal law of gravitation had been an inverse cubic law, rather than our familiar square law, a possible orbit of the planets around the Sun would have been that of a logarithmic spiral.

Logarithmic spirals are remarkable curves for very many reasons, and one will prove to be the second curve needed for the solution to our principal problem.

Before we do so, it is impossible to ignore an amusing anecdote relating to the remarkable analytic number theorist G. H. Hardy, in which he posits an equation which represents an equiangular spiral, which is also a parabola, and a hyperbola.

During his tenure of the Savilian Chair of Geometry at Oxford, he gave his presidential address to the Mathematical Association in 1925, under the title, 'What is geometry?', in the course of which he said with characteristic clarity:

> You might object...that geometry is, after all, the business of geometers, and that I know, and you know, and I know that you know, that I am not one; and that it is useless for me to try to tell you what geometry is, because I simply do not know. And here I am afraid that we are confronted with a regrettable but quite definite cleavage of opinion. I do not claim to know any geometry, but I do claim to understand quite clearly what geometry is.

He had, however, contributed to the geometrical literature with the following note, published in the *Mathematical Gazette* in 1907.

---

224. [M^1.8.g.] *A curious imaginary curve.*

The curve $(x + iy)^2 = \lambda(x - iy)$ is (i) a parabola, (ii) a rectangular hyperbola, and (iii) an equiangular spiral. The first two statements are evidently true. The polar equation is

$$r = \lambda e^{-3i\theta}$$

the equation of an equiangular spiral. The intrinsic equation is easily found to be $\rho = 3is$.

It is instructive (i) to show that the equation of any curve which is both a parabola and a rectangular hyperbola can be put in the form given above, or in the form

$$(x + iy)^2 = x \quad (\text{or } y)$$

and (ii) to determine the intrinsic equation directly from one of the latter forms of the Cartesian equation.

G. H. HARDY

We will not pursue his argument fully, partly as we have no wish to delve much into complex numbers here, but the 'evidently true' part of the statement seems to rely on two substitutions of variables:

- $X = x - iy$ and $Y = x + iy$, which transforms the equation to $Y^2 = \lambda X$, a parabola;
- $X = (x + iy)/(x - iy)$ and $Y = x + iy$, which transforms the equation to $XY = \lambda$, a rectangular hyperbola.

That it is also an equiangular spiral is shown by recourse to the polar form of complex numbers, $z = re^{i\theta}$. The equation may be written in the form $z^2 = \lambda z^*$, where z^* is the complex conjugate of z, the polar form of which is $z^* = re^{-i\theta}$. His equation then becomes

$$(re^{i\theta})^2 = \lambda(re^{-i\theta}),$$

which simplifies to

$$r = \lambda e^{-3i\theta}$$

and the equiangular spiral has resurfaced once more!
  Now to our central problem.

Our Pursuit Problem

In these earlier examples the pursuer is aware not only of the
speed of the pursuer but also the direction in which he is trav-
elling; what happens if we remove this second piece of intelligence? Doing so brings us to our problem:

> A smuggler, travelling as fast as possible in a straight line,
> is being pursued and caught up by a coastguard when a fog-
> bank engulfs them and each becomes invisible to the other.
> The smuggler's boat is too small for electronic detection or
> to leave an appreciable wake to follow, yet in spite of the
> coastguard not knowing where the smuggler is or in which
> direction he is then travelling, he can steer a course that
> guarantees capture of the smuggler.

One of the crucial words here is *guarantees*. This is not a matter
of luck, not a question of probability; the smuggler will be caught
with certainty, and with the use of a combination of that Circle
of Apollonius and that logarithmic spiral.

The Solution

First we employ the Circle of Apollonius. Writing the speed of
the coastguard as v_c and that of the smuggler as v_s and (for
convenience) $k = v_c/v_s > 1$, we construct figure 10.7 in the
following way.
  Assume that the smuggler enters the fogbank at P_1 and that
the coastguard is then at S_1. Define the number d by the condition that the distance $S_1 P_1 = (k + 1)d$ and construct the Circle
of Apollonius C_1 of all points A so that $AS_1 = kAP_1$ and let it
cross $S_1 P_1$ at the point Y; it is the case, then, that $S_1 Y = kd$ and
$YP_1 = d$. The captain of the coastguard vessel could aim for any

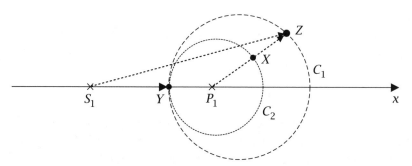

Figure 10.7. The solution diagram.

point on C_1 and, if he is lucky enough for the pirate to have chosen the appropriate direction, the interception will take place at Z, as we saw in the earlier example.

Now consider a different coastguard strategy: travel from S_1 to Y (a distance of kd) and so be a distance d from P_1. Since $v_s = (1/k)v_c$, the smuggler will have travelled a distance d and therefore be at some point X, somewhere on the circle C_2, which has centre P_1 and radius d. At this time both the smuggler and the coastguard are distant d from P_1. The plan now will be for the coastguard to maintain the same distance as the smuggler from P_1 and for their paths to cross. With P_1 as the origin and relative to the positive x-direction, write the polar equation of the coastguard's subsequent path as $r = r(\theta)$, with θ measured anticlockwise from that direction. We will ensure that they both maintain the same distance from P_1 if we make $dr/dt = v_s$. If s is the distance the coastguard has travelled from Y, then $ds/dt = v_c$, which makes

$$\frac{ds}{dr}\frac{dr}{dt} = v_c \quad \text{and} \quad \frac{ds}{dr} = \frac{v_c}{v_s} = k.$$

Now we use a standard result of calculus, the detail of which is given in appendix C,

$$\left(\frac{ds}{dr}\right)^2 = 1 + \left(r\frac{d\theta}{dr}\right)^2,$$

which means that

$$\left(r\frac{d\theta}{dr}\right)^2 = k^2 - 1 \quad \text{and} \quad r\frac{d\theta}{dr} = \sqrt{k^2 - 1}.$$

Rewritten as a standard integral, this becomes

$$\int d\theta = \sqrt{k^2 - 1} \int \frac{1}{r} \, dr$$

the solution to which is

$$r = ae^{\theta/\sqrt{k^2-1}}$$

for some constant a, and we have the polar equation of the logarithmic spiral.

We also have the condition that $r = d$ when $\theta = \pi$ and we can use this to evaluate the constant a. The equation

$$r = ae^{\theta/\sqrt{k^2-1}}$$

gives

$$d = ae^{\pi/\sqrt{k^2-1}}, \quad a = de^{-\pi/\sqrt{k^2-1}},$$

and so the equation of the spiral becomes

$$r = de^{\pi/\sqrt{k^2-1}}e^{\theta/\sqrt{k^2-1}} = de^{(\theta-\pi)/\sqrt{k^2-1}}$$

and the solution path to the problem is

$$r = de^{(\theta-\pi)/\sqrt{k^2-1}}.$$

This spiral must at some time cross the smuggler's path and when it does the two ships will be equidistant from P_1, as shown in figure 10.8, and so they will be in the same place. Capture is certain!

If the smuggler's track makes an angle φ with the positive x-direction, the capture will take place when $\theta = 2\pi + \varphi$ (where $-\pi \leqslant \theta \leqslant \pi$) and so we can calculate the distance travelled to interception by calculating the arc length of the path from $\theta = \pi$

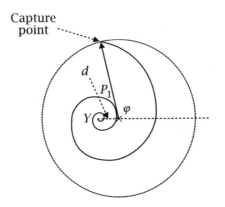

Figure 10.8. Captured!

to $\theta = 2\pi + \varphi$, which we can deduce from the second form of the result from appendix C,

$$\left(\frac{ds}{d\theta}\right)^2 = \left(\frac{dr}{d\theta}\right)^2 + r^2,$$

to get

$$s = \int_{\pi}^{2\pi+\varphi} \sqrt{\left(\frac{dr}{d\theta}\right)^2 + r^2}\, d\theta$$

$$= \int_{\pi}^{2\pi+\varphi} \sqrt{\left(\left(\frac{d}{\sqrt{k^2-1}}\right)^2 + d^2\right)e^{2(\theta-\pi)/\sqrt{k^2-1}}}\, d\theta$$

$$= \frac{dk}{\sqrt{k^2-1}} \int_{\pi}^{2\pi+\varphi} e^{(\theta-\pi)/\sqrt{k^2-1}}\, d\theta$$

$$= \frac{dk}{\sqrt{k^2-1}} [\sqrt{k^2-1}\, e^{(\theta-\pi)/\sqrt{k^2-1}}]_{\pi}^{2\pi+\varphi}$$

$$= dk(e^{(\pi+\varphi)/\sqrt{k^2-1}} - 1)$$

and the time to interception is given by

$$de^{(\theta-\pi)/\sqrt{k^2-1}} = v_s t,$$

where $\theta = 2\pi + \varphi$. This means that

$$t = (d/v_s)e^{(\pi+\varphi)/\sqrt{k^2-1}}.$$

Chapter 11

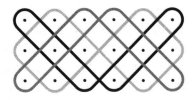

PARRONDO'S GAMES

This is a one line proof... if we start sufficiently far to the left.

Unknown Cambridge University lecturer

We are all used to the idea of losing in any number of games of chance, particularly in games which are biased against us. If we decide to alleviate the monotony of giving our money away by varying play between two such games, we would reasonably expect no surprises in the inevitability of our financial decline, but that is to ignore the discovery of Dr Juan Parrondo: *two losing games can combine to a winning composite game.*

We will avoid the rigorous definition of a losing game since we all have an instinctive feel for what one is, and that is enough for our purposes; in short, if we are foolish to gamble on such a game, in the long term we expect to finish up with less money than we started with. Put more mathematically, the probability of our winning on any play of the game is less than 0.5. (We can also lose even if it equals 0.5 if our wealth is small compared with the other player; the reader may wish to investigate the implications of what is known as *gambler's ruin*.) Now suppose that there are two such games, call them A and B, which we might play individually or in combination, with the pattern for playing the combination quite arbitrary: we could play A for a while and then change and continue gambling by playing B, or we could alternate playing ABABAB... or toss a (possibly biased) coin to determine which we play and when, etc. Whatever the strategy we use to decide which game to play and when, we would expect to lose in the long term, but the force of Parrondo's result is that situations exist which result in a winning combination. The public announcement was made (in particular) in the 1999 article, 'Parrondo's Paradox', by G. P. Harmer and D. Abbott in *Statistical Science* 14(2):206–13. Three years earlier the Spanish physicist Juan M. R. Parrondo had presented the idea in unpublished form at a workshop in Turin, Italy: he had defined a composite, winning game from two provably losing games; that is, the player's fortune provably increases as he continues to play the combined game. It is this process which attracts our attention in this chapter, as we look at *Parrondo's Games*.

The Basic Game

The study is part of the much bigger topic of Markov chains, but here we need the simplest of ideas from them. Suppose that we repeatedly play a game and either win 1 unit with a constant probability p or lose 1 unit with probability $1 - p$. We will start with some fortune and write P_r for the probability that our fortune is reduced to 0 when it is currently at a level r; we will not consider the possibility of exhausting the wealth of our opponent.

Figure 11.1 summarizes the position as we take one more gamble, which means that

$$P_r = pP_{r+1} + (1 - p)P_{r-1}, \quad r \geqslant 1, \tag{1}$$

and also we clearly have that $P_0 = 1$.

This summarizes such a game and provides us with what is known as a recurrence relation for the P_r; what we would like is an explicit expression for P_r in terms of p and r.

With a starting fortune of r and
a consequent probability P_r that
our fortune is reduced to 0

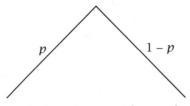

p $1 - p$

A win and so our fortune is   A loss and so our fortune is
now $r + 1$ and the probability   now $r - 1$ and the probability
that our fortune is reduced   that our fortune is reduced
to 0 is P_{r+1} to 0 is P_{r-1}

Figure 11.1. The basic tree diagram.

The trick that accomplishes this is to try a solution of the form $P_r = x^r$, which results in

$$x^r = px^{r+1} + (1 - p)x^{r-1}$$

and cancelling by x^{r-1} results in $x = px^2 + (1 - p)$ or

$$px^2 - x + (1 - p) = 0.$$

This quadratic equation factorizes to $(x - 1)(px - (1 - p)) = 0$ and so has roots $x = 1$ and $x = (1 - p)/p$. This means that

$$P_r = 1^r = 1 \quad \text{and} \quad P_r = \left(\frac{1-p}{p}\right)^r$$

are both solutions, which we can easily check by substituting each of them in equation (1).

This isn't the full story, though. Again, it is easy to check that any constant multiple of each solution is again a solution, and, further, that the sum of each of the two solutions is itself a solution. This results in the completely general solution of (1) of

$$P_r = P\left(\frac{1-p}{p}\right)^r + Q, \quad r \geq 0,$$

for constants P and Q.

We can now invoke the condition that $P_0 = 1$ to get $P + Q = 1$, which provides us with one equation in two unknowns and to find unique values of P and Q we need a second, independent equation. This is not quite so easy.

If the opponent had a known capital, which would allow us to write the total initial capital between the two players as, say, N, we could also invoke the condition $P_N = 0$, which would result in that second equation for P and Q and this would mean that the values could be established. As it is, we have no such condition, but we can make progress by using the following intuitive argument.

It must be that $(1 - p)/p$ is equal to, greater than or less than 1, and we can consider the three cases separately.

If $(1 - p)/p = 1$ (that is, $p = \frac{1}{2}$), then $P_r = P + Q = 1$ and in the long term we are assured of losing all of the capital against an opponent of much greater wealth.

If $(1-p)/p > 1$, as r increases P_r is bound to fall outside $[0, 1]$ and so violate the laws of probability, that is, unless $P = 0$, which makes $P_r = 1$ for all r.

Now suppose that $(1 - p)/p < 1$, then, as r increases,

$$\left(\frac{1 - p}{p}\right)^r \to 0$$

but as we approach an infinite resource our probability of losing must approach 0, which means that $Q = 0$ and so $P = 1$. In this case,

$$P_r = \left(\frac{1 - p}{p}\right)^r \quad \text{for } r \geqslant 0$$

and we conclude that

$$P_r = \begin{cases} 1, & \dfrac{1 - p}{p} \geqslant 1 \quad \text{(losing)}, \\[2mm] \left(\dfrac{1 - p}{p}\right)^r, & \dfrac{1 - p}{p} < 1 \quad \text{(winning)}. \end{cases} \tag{2}$$

Of course, the inequalities for p simplify to $p \leqslant \frac{1}{2}$ and $p > \frac{1}{2}$ and this accords with our intuitive idea that if the game is fair

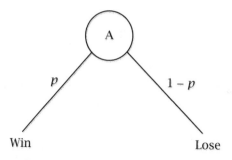

Figure 11.2. Game A tree diagram.

or biased against us, eventually we will probably lose all of our fortune. As we mentioned before, even a fair game is a losing game if we have no chance of exhausting the opponent's fortune.

The Parrondo Setup

This analysis is repeated three times to expose the Parrondo paradox: there will be a game A and another game B, both of which are losing. We will then combine them, to result in a multiple game C, which will prove to be winning.

Game A

This is simply the game we have already described. Think of it as tossing a biased coin and getting a winning head with a fixed probability p or a losing tail with probability $1 - p$. The winning and losing conditions are then represented by the same diagram as before, which is repeated in shortened form in figure 11.2. The solution is provided by equation (2) on page 118:

$$
P_r = \begin{cases} 1, & p \leqslant \frac{1}{2} \quad \text{(losing)}, \\[2mm] \left(\dfrac{1-p}{p}\right)^r, & p > \frac{1}{2} \quad \text{(winning)}. \end{cases}
$$

Game B

This is more complicated, with the chance of winning on any play of it dependent on the size of the capital at the time. To be exact, if the capital happens to be a multiple of 3, we win with

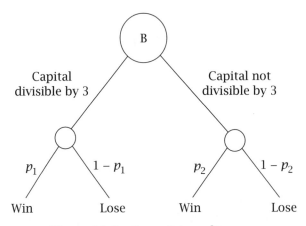

Figure 11.3. Game B tree diagram.

probability p_1; otherwise, we win with probability p_2, summed up in figure 11.3.

Since any positive integer must be one of $\{3r, 3r+1, 3r+2\}$, we must consider three recurrence relations, the first generated by the left branch of the diagram and the remainder by its right side.

With P_r defined as before we then have

$$P_{3r} = p_1 P_{3r+1} + (1 - p_1)P_{3r-1}, \quad r \geqslant 1, \tag{3}$$

$$P_{3r+1} = p_2 P_{3r+2} + (1 - p_2)P_{3r}, \qquad r \geqslant 0, \tag{4}$$

$$P_{3r+2} = p_2 P_{3r+3} + (1 - p_2)P_{3r+1}, \quad r \geqslant 0, \tag{5}$$

together with the condition $P_0 = 1$.

Again, we want an explicit formula for P_r in terms of r and we work towards this by first finding a formula for P_{3r}.

The algebra is devious and we start by writing equations (4) and (5) as

$$P_{3r+1} - p_2 P_{3r+2} = (1 - p_2)P_{3r}$$

and

$$P_{3r+2} - (1 - p_2)P_{3r+1} = p_2 P_{3r+3}$$

and so think of them as two equations in the two unknowns P_{3r+1} and P_{3r+2}.

These have the solutions (after some standard but messy algebra)

$$P_{3r+1} = \frac{P_{3r} - 2p_2 P_{3r} + p_2^2 P_{3r} - p_2^2 P_{3r+3}}{2p_2 - 1},$$

$$P_{3r+2} = \frac{P_{3r} - p_2 P_{3r} - p_2 P_{3r+3}}{2p_2 - 1}.$$

Rewriting P_{3r+2} with $r - 1$ replacing r then gives

$$P_{3r-1} = \frac{P_{3r-3} - p_2 P_{3r-3} - p_2 P_{3r}}{2p_2 - 1}$$

and substituting these expressions for P_{3r+1} and P_{3r-1} into equation (3) and simplifying gives

$$P_{3r}(1 - p_1 - 2p_2 + p_2^2 + 2p_1 p_2)$$
$$= p_1 p_2^2 P_{3r+3} + (1 - p_1)(1 - p_2)^2 P_{3r-3},$$

which can be rewritten as

$$P_{3r}((1 - p_1)(1 - p_2)^2 + p_1 p_2^2)$$
$$= p_1 p_2^2 P_{3r+3} + (1 - p_1)(1 - p_2)^2 P_{3r-3}$$

or

$$P_{3r} = \frac{p_1 p_2^2}{(1 - p_1)(1 - p_2)^2 + p_1 p_2^2} P_{3r+3}$$
$$+ \frac{(1 - p_1)(1 - p_2)^2}{(1 - p_1)(1 - p_2)^2 + p_1 p_2^2} P_{3r-3}.$$

This may be messy but a careful look reveals that the sum of the coefficients is 1 and therefore that this has the form of equation (1) with

$$p = \frac{p_1 p_2^2}{(1 - p_1)(1 - p_2)^2 + p_1 p_2^2}$$

and r replaced by $3r$.

We then have that

$$\frac{1 - p}{p} = \frac{(1 - p_1)(1 - p_2)^2}{p_1 p_2^2}$$

and so

$$P_{3r} = A\left(\frac{(1-p_1)(1-p_2)^2}{p_1 p_2^2}\right)^{3r} + B.$$

Exactly the same arguments as before, only this time with

$$\frac{(1-p_1)(1-p_2)^2}{p_1 p_2^2},$$

result in

$$P_{3r} = \begin{cases} 1, & \dfrac{(1-p_1)(1-p_2)^2}{p_1 p_2^2} \geqslant 1 \\[4mm] & \text{(losing)}, \\[4mm] \left(\dfrac{(1-p_1)(1-p_2)^2}{p_1 p_2^2}\right)^{3r}, & \dfrac{(1-p_1)(1-p_2)^2}{p_1 p_2^2} < 1 \\[4mm] & \text{(winning)}. \end{cases}$$

Repeat the same (rather tedious) working with P_{3r+1} and P_{3r+2} and we would get the equivalent expressions with $3r$ replaced by $3r + 1$ and $3r - 1$, respectively, and we can compress the whole thing into what we have been seeking: an explicit expression for P_r in terms of r, which is

$$P_r = \begin{cases} 1, & \dfrac{(1-p_1)(1-p_2)^2}{p_1 p_2^2} \geqslant 1 \\[4mm] & \text{(losing)}, \\[4mm] \left(\dfrac{(1-p_1)(1-p_2)^2}{p_1 p_2^2}\right)^{r}, & \dfrac{(1-p_1)(1-p_2)^2}{p_1 p_2^2} < 1 \\[4mm] & \text{(winning)}. \end{cases}$$

Notice that if we put $p_1 = p_2 = p$, the expression is exactly that of game A: game A is indeed a special case of game B.

Now for the payoff, as we consider the composite game C.

Game C

Suppose that we play game A with a probability of y and game B with a probability of $1 - y$. All depends on whether or not our current capital is a multiple of 3, as shown figure 11.4.

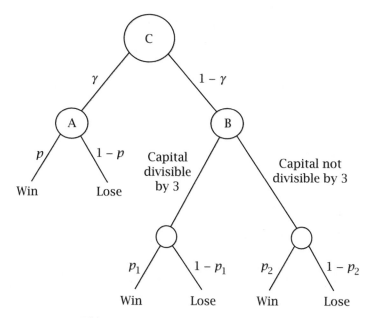

Figure 11.4. Game C tree diagram.

If our capital is a multiple of 3, we will win with a probability $q_1 = yp + (1 - y)p_1$ and otherwise we win with a probability $q_2 = yp + (1 - y)p_2$. This is an identical setup to game B with q replacing p, which means that

$$
P_r = \begin{cases}
1, & \dfrac{(1 - q_1)(1 - q_2)^2}{q_1 q_2^2} \geqslant 1 \\
& \qquad \text{(losing),} \\
\left(\dfrac{(1 - q_1)(1 - q_2)^2}{q_1 q_2^2} \right)^r, & \dfrac{(1 - q_1)(1 - q_2)^2}{q_1 q_2^2} < 1 \\
& \qquad \text{(winning).}
\end{cases}
$$

We will have a paradoxical position if

$$
\frac{1 - p}{p} > 1, \qquad \text{game A losing,}
$$

$$
\frac{(1 - p_1)(1 - p_2)^2}{p_1 p_2^2} > 1, \qquad \text{game B losing,}
$$

$$
\frac{(1 - q_1)(1 - q_2)^2}{q_1 q_2^2} < 1, \qquad \text{game C winning.}
$$

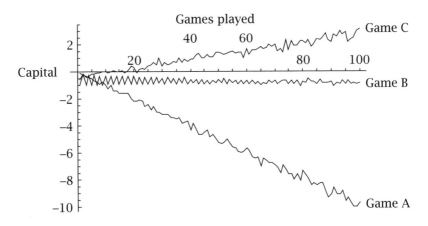

Figure 11.5. Parrondo's Paradox.

We could take, for example, $p = 0.45$, $p_1 = 0.01$, $p_2 = 0.90$, $y = 0.5$ to give

$$\frac{1-p}{p} = 1.\dot{2} > 1,$$

$$\frac{(1-p_1)(1-p_2)^2}{p_1 p_2^2} = 1.\dot{2} > 1,$$

$$\frac{(1-q_1)(1-q_2)^2}{q_1 q_2^2} = 0.776\ldots < 1.$$

A simulation of the play of the games is given in figure 11.5. The plot is the average fortune over 1000 trials when each of game A, B and C is played up to 100 times with the probabilities given above.

Exactly What Is Happening?

The choices for the probabilities may seem arbitrary (and to some extent they are), but we can see further into the matter if we look at a plot of the function

$$\frac{(1-p_1)(1-p_2)^2}{p_1 p_2^2} = 1.$$

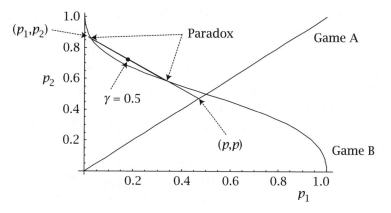

Figure 11.6. The paradox exposed.

If we consider this as p_2 being a function of p_1 and write that function explicitly, we have

$$p_2 = \frac{p_1 - 1 \pm \sqrt{p_1(1 - p_1)}}{2p_1 - 1} \quad \text{for } 0 \leqslant p_1 \leqslant 1$$

and since $0 \leqslant p_2 \leqslant 1$ we are interested only in

$$p_2 = \frac{p_1 - 1 + \sqrt{p_1(1 - p_1)}}{2p_1 - 1},$$

which is the curve in figure 11.6. Its behaviour at $p_1 = 0.5$ is defined by continuity.

Points on and below the curve represent pairs p_1, p_2 for which game B is losing. The diagonal straight line is $p_2 = p_1$ and so points on it represent choices of p in game A; below and on the intersection point $(0.5, 0.5)$ game A is losing and above it game A is winning.

Now recall the probabilities

$$q_1 = yp + (1 - y)p_1 \quad \text{and} \quad q_2 = yp + (1 - y)p_2$$

of game C winning. If we write these equations as

$$\begin{pmatrix} q_1 \\ q_2 \end{pmatrix} = y \begin{pmatrix} p \\ p \end{pmatrix} + (1 - y) \begin{pmatrix} p_1 \\ p_2 \end{pmatrix}$$

and let y vary, we can see that the pairs (q_1, q_2) lie on the straight line joining (p, p) to (p_1, p_2). The paradox exists when the points (p, p) and (p_1, p_2) are chosen so that the line passes above the curve, since game B for these values (that is, game C) is winning. It is the convexity of the curve that allows the paradox to exist. The third line joins the two points defined by our choice of probabilities above and marked on it is the point corresponding to our choice of $y = 0.5$.

It may be a little contrived, but here we have a procedure for making a winning game out of two losing games. Since the idea was introduced, there have been many examples of real-world manifestations of the fact that a combination of two negative characteristics can result in a positive one. To demonstrate the diversity that exists, in 2000 in the *New York Times*, Dr Sergei Maslov from Brookhaven National Laboratory was reported to have shown that if an investor simultaneously shared capital between two losing stock portfolios, capital would increase rather than decrease; Brooke Buckley, an undergraduate student from Eastern Kentucky University, mentions in her thesis the well-known fact in agriculture 'that both sparrows and insects can eat all the crops. However, by having a combination of sparrows and insects, a healthy crop is harvested.'

Chapter 12

HYPERDIMENSIONS

I recall a lecture by John Glenn, the first American to go into orbit. When asked what went through his mind while he was crouched in the rocket nose-cone, awaiting blast-off, he replied, 'I was thinking that the rocket has 20,000 components, and each was made by the lowest bidder.'

Martin Rees

Some dimensionally dependent phenomena seem reasonable. Take, for instance, the idea of a random walk. In one dimension this means that we start at the origin and move to the left or the right with equal probability; in two dimensions we have four equally probable directions in which we can walk; in both cases it can be shown that the probability of eventually returning to the origin is 1; we cannot, in theory, get lost. As the dimension increases, so we might reasonably think are the chances of getting lost, never to return to the origin, increase, and so they do. In three dimensions the probability of our return is only about 0.34 and in n dimensions, where n is large, that probability is about $1/2n$. But it does not take much investigation into hyperdimensions to cause our intuition to be confounded.

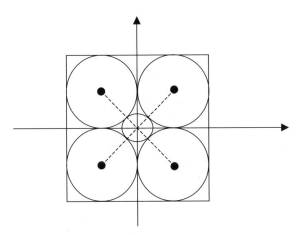

Figure 12.1. Touching circles.

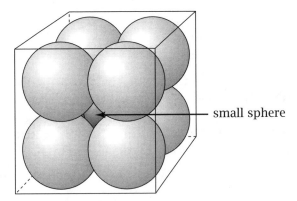

Figure 12.2. Touching spheres.

A Strange Phenomenon

Figure 12.1 shows a 4×4 square inside of which is embedded four touching circles, each of radius 1. It also shows a smaller fifth circle of radius $\sqrt{1^2 + 1^2} - 1 = \sqrt{2} - 1$ inscribed at the centre of the square, itself touching all four circles. There is nothing surprising here.

Figure 12.2 shows the equivalent situation in three dimensions. Inside a cube of side 4 are embedded eight touching spheres, each of radius 1, and once again there is room for a smaller, extra sphere which touches all eight, this time its radius is $\sqrt{1^2 + 1^2 + 1^2} - 1 = \sqrt{3} - 1$. Again, this is self-evident.

It is also self-evident that in both cases the central circle or sphere is contained in the surrounding square or cube.

Now suppose that we move from two to three to n dimensions and consider an n-dimensional hypercube, in which hyperspheres are inscribed. The definitions of these objects are reasonable enough.

An n-dimensional hypercube of side L (with one vertex at the origin) is the set of all n-tuples $\{x_1, x_2, x_3, \ldots, x_n\}$, where $x_r \in \{0, L\}$: the figure has 2^n vertices (and we can embed $\binom{N}{n} 2^{N-n}$ of them in $(N \geqslant n)$-dimensional space).

An n-dimensional hypersphere of radius R is the set of all n-tuples $\{x_1, x_2, x_3, \ldots, x_n\}$ such that $x_1^2 + x_2^2 + x_3^2 + \cdots + x_n^2 \leqslant R^2$.

Currently, we are interested in the case $L = 4$ and if we carry out the natural calculation for the radius of the small, inscribed hypersphere using the generalized Pythagoras Theorem, we get

$$r = \underbrace{\sqrt{1^2 + 1^2 + 1^2 + \cdots + 1^2}}_{n \text{ times}} - 1 = \sqrt{n} - 1.$$

Putting $n = 2$ or 3 gives us the previous results, but reflect on the fact that the distance from the centre of the hypercube to any of its sides is always precisely 2 units. Now consider the situation when $n = 9$; the radius of the inner, touching hypersphere $r = \sqrt{9} - 1 = 2$, which must mean that it touches the sides of the hypercube, and, when $n > 9$, it protrudes outside it!

Furthermore, using Pythagoras's Theorem on an n-dimensional hypercube of side m, the length of the 'space diagonal' is

$$r = \underbrace{\sqrt{m^2 + m^2 + \cdots + m^2}}_{n \text{ times}} = m\sqrt{n},$$

so if we wanted to place a stick of length L into the hypercube we would require $m\sqrt{n} = L$, which means that, as the dimension increases, the side of the hypercube needed to contain the stick diminishes. For example, a hypercube of side 1 metre of dimension 100 could contain a stick of length 10 metres; one of dimension 2.25×10^6 could contain a stick of length the metric mile (1500 metres).

Extrapolating from what is evident from our three-dimensional perspective to what is logically, but uncomfortably, true in hyperspace forms the substance of this chapter, and also of a famous piece of mathematical literature.

The Literary Dimension

The little Hexagon meditated on this a while and then said to me:

'But you have been teaching me to raise numbers to the third power: I suppose three-to-the-third must mean something in Geometry; what does it mean?'

'Nothing at all,' replied I, 'not at least in Geometry; for Geometry has only Two Dimensions.'

And then I began to show the boy how a Point by moving through a length of three inches makes a Line of three inches, which may be represented by three; and how a Line of three inches, moving parallel to itself through a length of three inches, makes a Square of three inches every way, which may be represented by three-to-the-second. Upon this, my Grandson, again returning to his former suggestion, took me up rather suddenly and exclaimed,

'Well, then, if a Point by moving three inches, makes a Line of three inches represented by three; and if a straight Line of three inches, moving parallel to itself, makes a Square of three inches every way, represented by three-to-the-second; it must be that a Square of three inches every way, moving somehow parallel to itself (but I don't see how) must make something else (but I don't see what) of three inches every way – and this must be represented by three-to-the-third.'

The dialogue originates from a conversation between a Square, the principal character of Edwin A. Abbott's 1884 mathematical romance, *Flatland* (http://www.gutenberg.org/dirs/etext94/flat11.txt), and his gifted, regular-hexagonal grandson. In a dream the Square, an inhabitant of a two-dimensional world, had already failed to explain Flatland to the monarch of one-dimensional Lineland; now his grandson had challenged him to comprehend a dimension above that in which they both lived.

A little later, a stranger from Spaceland appeared first as a point, which became a small circle, and which grew continuously to a circle of greatest size, which diminished to a point and which disappeared altogether; incomprehensible to the Flatlanders, that stranger was a three-dimensional sphere which had passed through their two-dimensional world.

The earlier examples provided an elementary case in which our Spacelander perception is confounded and the remainder of this chapter concentrates on several other hyperdimensional results, which are at the very least exotic.

Volumes in Discrete Hyperdimensions

The volume of a hypercube of side L is very straightforward to calculate: $C_n(L) = L^n$. In particular, we have that

$$\begin{array}{c} C_n(L) \\ _{n \to \infty} \end{array} \begin{cases} \to \infty, & L > 1, \\ = 1, & L = 1, \\ \to 0, & L < 1. \end{cases}$$

Now we will undertake the significantly greater challenge of finding $V_n(X)$, the volume of the n-dimensional hypersphere of radius X.

It is important to understand that this is related to the volume of the n-dimensional hypersphere of radius 1 by $V_n(X) = X^n V_n(1)$, since the transition from a hypersphere of radius 1 to one of radius X may be thought of as a uniform change of units in each of the dimensions; the rules of similarity dictate that the volume will change by the product of these changes.

To approach the problem of finding $V_n(X)$ we could naturally delve into the world of n-fold multiple integrals, but it is wise to avoid such esotericism. Instead, we will extend the use of Cavalieri's Principle (chapter 8) from finding the volume of a sphere to that of our hypersphere.

The principle tells us that to find the volume of a solid, take an arbitrary (x) axis which runs through it and add the areas of the sections $A(x)$ through the solid, perpendicular to this axis, as demonstrated in figure 12.3.

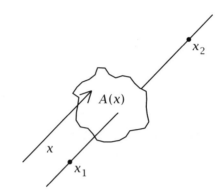

Figure 12.3. Finding volume using Cavalieri's Principle.

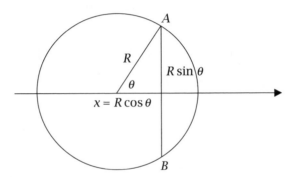

Figure 12.4. A section of a three-dimensional sphere.

In symbols, the volume is $\int_{x_1}^{x_2} A(x)\,\mathrm{d}x$. Now apply this to our three-dimensional sphere, centred at the origin, and take the x-axis as the chosen axis.

Figure 12.4 gives a sectional view of the sphere with $A(x)$ a circular section distant $x = R\cos\theta$ from its centre and of diameter AB; the radius of the cross-section is therefore $R\sin\theta$. We then have

$$V_3(R) = \int_{x_1}^{x_2} A(x)\,\mathrm{d}x,$$

where

$$A(x) = V_2(R\sin\theta) = (R\sin\theta)^2 V_2(1).$$

Since $x = R\cos\theta$ we have $\mathrm{d}x/\mathrm{d}\theta = -R\sin\theta$ and can substitute

θ for x in the integral to get

$$V_3(R) = \int_{-R}^{R} (R \sin \theta)^2 V_2(1) \, dx$$

$$= \int_{\pi}^{0} (R \sin \theta)^2 V_2(1) \times -R \sin \theta \, d\theta$$

$$= R^3 V_2(1) \int_{0}^{\pi} \sin^3 \theta \, d\theta$$

and the problem reduces to finding a standard integral.

The calculation continues as

$$V_3(R) = R^3 V_2(1) \int_{0}^{\pi} \sin \theta \sin^2 \theta \, d\theta$$

$$= R^3 V_2(1) \int_{0}^{\pi} \sin \theta (1 - \cos^2 \theta) \, d\theta$$

$$= R^3 V_2(1) \int_{0}^{\pi} \sin \theta - \sin \theta \cos^2 \theta \, d\theta$$

$$= R^3 V_2(1) [- \cos \theta + \tfrac{1}{3} \cos^3 \theta]_0^{\pi}$$

$$= R^3 V_2(1)(\tfrac{2}{3} + \tfrac{2}{3}) = \tfrac{4}{3} R^3 V_2(1) = \tfrac{4}{3} \pi R^3$$

noting that $V_2(1) = \pi \times 1^2 = \pi$.

Having established a comfortingly familiar result, we will continue to find a general formula for the volume of the n-dimensional hypersphere, $V_n(R)$, noting that $A(x)$ will be the volume of an $(n-1)$-dimensional hypersphere.

Following the same route, we get

$$V_n(R) = \int_{x_1}^{x_2} V_{n-1}(x) \, dx$$

$$= \int_{-R}^{R} V_{n-1}(x) \, dx$$

$$= \int_{\pi}^{0} V_{n-1}(R \sin \theta) \times -R \sin \theta \, d\theta$$

$$= \int_{\pi}^{0} (R \sin \theta)^{n-1} V_{n-1}(1) \times -R \sin \theta \, d\theta$$

$$= V_{n-1}(1) R^n \int_{0}^{\pi} \sin^n \theta \, d\theta.$$

So

$$V_n(R) = R^n V_n(1) = V_{n-1}(1)R^n \int_0^\pi \sin^n \theta \, d\theta$$

and we conclude that

$$V_n(1) = V_{n-1}(1) \int_0^\pi \sin^n \theta \, d\theta = V_{n-1}(1)I_n.$$

We have a recursive formula for $V_n(1)$ which also involves the integral I_n and to derive an explicit formula we will first find an explicit formula for I_n and use it to chase down the $V_n(1)$.

First, the I_n, which we attack in a standard way, using integration by parts to establish a reduction formula:

$$\begin{aligned}
I_n &= \int_0^\pi \sin^n \theta \, d\theta = \int_0^\pi \sin \theta \times \sin^{n-1} \theta \, d\theta \\
&= [-\cos \theta \times \sin^{n-1} \theta]_0^\pi + (n-1) \int_0^\pi \cos^2 \theta \times \sin^{n-2} \theta \, d\theta \\
&= (n-1) \int_0^\pi (1 - \sin^2 \theta) \times \sin^{n-2} \theta \, d\theta \\
&= (n-1) \int_0^\pi \sin^{n-2} \theta - \sin^n \theta \, d\theta \\
&= (n-1)I_{n-2} - (n-1)I_n.
\end{aligned}$$

This means that $I_n = ((n-1)/n)I_{n-2}$, which we can use to find explicit formulae for I_n, depending on whether n is even or odd.

For n even:

$$\begin{aligned}
I_n &= \frac{n-1}{n} I_{n-2} = \frac{n-1}{n} \frac{n-3}{n-2} I_{n-4} \\
&= \frac{n-1}{n} \frac{n-3}{n-2} \frac{n-5}{n-4} I_{n-6} \\
&= \frac{n-1}{n} \frac{n-3}{n-2} \frac{n-5}{n-4} \cdots \frac{3}{4} \frac{1}{2} I_0 \\
&= \frac{n-1}{n} \frac{n-3}{n-2} \frac{n-5}{n-4} \cdots \frac{3}{4} \frac{1}{2} \int_0^\pi 1 \, d\theta \\
&= \frac{n-1}{n} \frac{n-3}{n-2} \frac{n-5}{n-4} \cdots \frac{3}{4} \frac{1}{2} \pi.
\end{aligned}$$

For n odd:

$$
\begin{aligned}
I_n &= \frac{n-1}{n} I_{n-2} = \frac{n-1}{n} \frac{n-3}{n-2} I_{n-4} \\
&= \frac{n-1}{n} \frac{n-3}{n-2} \frac{n-5}{n-4} I_{n-6} \\
&= \frac{n-1}{n} \frac{n-3}{n-2} \frac{n-5}{n-4} \cdots \frac{4}{5} \frac{2}{3} I_1 \\
&= \frac{n-1}{n} \frac{n-3}{n-2} \frac{n-5}{n-4} \cdots \frac{4}{5} \frac{2}{3} \int_0^\pi \sin\theta \, d\theta \\
&= \frac{n-1}{n} \frac{n-3}{n-2} \frac{n-5}{n-4} \cdots \frac{4}{5} \frac{2}{3} 2.
\end{aligned}
$$

Now we use these to establish a connection between I_n and I_{n-1} that is independent of the parity of n.

If n is even, it must be that $n - 1$ is odd and we use the appropriate formulae to get

$$
\begin{aligned}
I_n I_{n-1} &= \frac{n-1}{n} \frac{n-3}{n-2} \frac{n-5}{n-4} \cdots \frac{3}{4} \frac{1}{2} \pi \\
&\times \frac{n-2}{n-1} \frac{n-4}{n-3} \frac{n-6}{n-5} \cdots \frac{4}{5} \frac{2}{3} 2 \\
&= \frac{2\pi}{n}.
\end{aligned}
$$

If n is odd, it must be that $n - 1$ is even and again we use the appropriate formulae to get

$$
\begin{aligned}
I_n I_{n-1} &= \frac{n-1}{n} \frac{n-3}{n-2} \frac{n-5}{n-4} \cdots \frac{4}{5} \frac{2}{3} 2 \\
&\times \frac{n-2}{n-1} \frac{n-4}{n-3} \frac{n-6}{n-5} \cdots \frac{3}{4} \frac{1}{2} \pi \\
&= \frac{2\pi}{n}.
\end{aligned}
$$

So, whatever the parity of n, $I_n I_{n-1} = 2\pi/n$.

Now recall that $V_n(1) = V_{n-1}(1) I_n$ and reuse the formula once on itself to get

$$
V_n(1) = V_{n-1}(1) I_n = (V_{n-2}(1) I_{n-1}) I_n = V_{n-2}(1)(I_{n-1} I_n).
$$

Table 12.1. Volumes of hyperspheres.

n	$V_n(R)$	$V_n(1)$
2	πR^2	$\pi = 3.141\,59\ldots$
3	$\frac{4}{3}\pi R^3$	$\frac{4}{3}\pi = 4.188\,79\ldots$
4	$\frac{1}{2}\pi^2 R^4$	$\frac{1}{2}\pi^2 = 4.934\,8\ldots$
5	$\frac{8}{15}\pi^2 R^5$	$\frac{8}{15}\pi^2 = 5.263\,7\ldots$
6	$\frac{1}{6}\pi^3 R^6$	$\frac{1}{6}\pi^3 = 5.167\,71\ldots$
7	$\frac{16}{105}\pi^3 R^7$	$\frac{16}{105}\pi^3 = 4.724\,76\ldots$
8	$\frac{1}{24}\pi^4 R^8$	$\frac{1}{24}\pi^4 = 4.058\,71\ldots$

Now we can say that

$$V_n(1) = V_{n-2}(1)(I_{n-1}I_n) = \frac{2\pi}{n}V_{n-2}(1).$$

We have a simple reduction formula for the $V_n(1)$ and we can chase it down, again depending on whether n is even or odd to get the answer:

$$V_n(1) = \begin{cases} \dfrac{2\pi}{n}\dfrac{2\pi}{n-2}\dfrac{2\pi}{n-4}\cdots\dfrac{2\pi}{2}1, & n \text{ even,} \\[2mm] \dfrac{2\pi}{n}\dfrac{2\pi}{n-2}\dfrac{2\pi}{n-4}\cdots\dfrac{2\pi}{3}2, & n \text{ odd.} \end{cases}$$

Of course, this means that

$$V_n(R) = R^n \begin{cases} \dfrac{2\pi}{n}\dfrac{2\pi}{n-2}\dfrac{2\pi}{n-4}\cdots\dfrac{2\pi}{2}1, & n \text{ even,} \\[2mm] \dfrac{2\pi}{n}\dfrac{2\pi}{n-2}\dfrac{2\pi}{n-4}\cdots\dfrac{2\pi}{3}2, & n \text{ odd.} \end{cases}$$

Table 12.1 lists the volumes for small values of n and we can see that the volume of the unit hypersphere peaks when $n = 5$ and that this maximal volume is $\frac{8}{15}\pi^2$. A plot of $V_n(1)$ against n is shown in figure 12.5, which indicates that the volume of the unit hypersphere is *decreasing* as n increases beyond 5, which appears very strange.

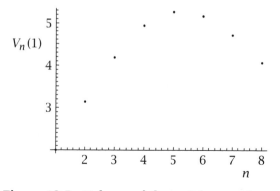

Figure 12.5. Volume of the unit hypersphere
compared with integer dimensions.

Volume in Continuous Hyperdimensions

The plotted points in figure 12.5 are in such a regular pattern
that it is natural to want to join them by a continuous curve, but
doing so would force us to admit not only hyperdimensions but
also nonintegral hyperdimensions. To approach this concept we
need to rewrite the formulae for $V_n(1)$ as below:

$$V_n(1) = \begin{cases} \dfrac{2\pi}{n}\dfrac{2\pi}{n-2}\dfrac{2\pi}{n-4}\cdots\dfrac{2\pi}{2}1, & n \text{ even}, \\[2ex] \dfrac{2\pi}{n}\dfrac{2\pi}{n-2}\dfrac{2\pi}{n-4}\cdots\dfrac{2\pi}{3}2, & n \text{ odd}, \end{cases}$$

$$= \begin{cases} \dfrac{\pi}{\frac{1}{2}n}\dfrac{\pi}{\frac{1}{2}n-1}\dfrac{\pi}{\frac{1}{2}n-2}\cdots\dfrac{\pi}{1}1, & n \text{ even}, \\[2ex] \dfrac{\pi}{\frac{1}{2}n}\dfrac{\pi}{\frac{1}{2}n-1}\dfrac{\pi}{\frac{1}{2}n-2}\cdots\dfrac{\pi}{\frac{3}{2}}2, & n \text{ odd}. \end{cases}$$

If n is even, the formula can be written in terms of factorials
as

$$V_n(1) = \frac{\pi^{n/2}}{(\frac{1}{2}n)!}.$$

If n is odd, $(\frac{1}{2}n)!$ is not defined, but its generalization – the
Gamma function $\Gamma(x)$ – is. This function's somewhat strange

definition is

$$\Gamma(x) = \int_0^\infty t^{x-1} e^{-t} \, dt,$$

which is defined for $x > 0$ and which has two particular properties:

$$\Gamma(1) = \int_0^\infty e^{-t} \, dt = [-e^{-t}]_0^\infty = 1$$

and

$$\Gamma(x+1) = \int_0^\infty t^x e^{-t} \, dt = [-t^x e^{-t}]_0^\infty + x \int_0^\infty t^{x-1} e^{-t} \, dt = x\Gamma(x).$$

Together they characterize the factorial function, since, if n is a positive integer,

$$\begin{aligned}
\Gamma(n) &= (n-1)\Gamma(n-1) = (n-1)(n-2)\Gamma(n-2) \\
&= (n-1)(n-2)(n-3)\Gamma(n-3) \\
&= (n-1)(n-2)(n-3) \cdots \Gamma(1) = (n-1)!.
\end{aligned}$$

So, this peculiar function is indeed an extension of the factorial function, which is defined only for positive integers, to all $x > 0$. In fact, the above relationship, when rewritten as

$$\Gamma(x) = \frac{1}{x}\Gamma(x+1),$$

can be used to extend the idea of factorial to all numbers other than the negative integers and it is perfectly easy to extend the definition to complex numbers, but we will not concern ourselves with these excitements.

Note in particular that, if we accept a standard result that

$$\int_0^\infty e^{-u^2} \, du = \frac{\sqrt{\pi}}{2},$$

we have

$$\Gamma(\tfrac{1}{2}) = \int_0^\infty t^{-1/2} e^{-t} \, dt$$

and using the substitution $t = u^2$, we have that $dt/du = 2u$ and

$$\int_0^\infty t^{-1/2}e^{-t}\,dt = \int_0^\infty u^{-1}e^{-u^2}2u\,du = 2\int_0^\infty e^{-u^2}\,du$$

and this means that

$$\Gamma(\tfrac{1}{2}) = 2\frac{\sqrt{\pi}}{2} = \sqrt{\pi}.$$

It is then the case that for n even $V_n(1)$ can be rewritten yet again, this time as

$$V_n(1) = \frac{\pi^{n/2}}{(\tfrac{1}{2}n)!} = \frac{\pi^{n/2}}{\Gamma(\tfrac{1}{2}n + 1)}.$$

The really nice thing is that this notation unifies the two formulae and it is easy to check that the Gamma function form of the formula holds whatever the parity of n.

For example,

$$V_5(1) = \frac{\pi^{5/2}}{\Gamma(\tfrac{5}{2} + 1)} = \frac{\pi^{5/2}}{\tfrac{5}{2}\Gamma(\tfrac{5}{2})} = \frac{\pi^{5/2}}{\tfrac{5}{2}\Gamma(\tfrac{3}{2} + 1)} = \frac{\pi^{5/2}}{\tfrac{5}{2}\tfrac{3}{2}\Gamma(\tfrac{3}{2})}$$

$$= \frac{\pi^{5/2}}{\tfrac{5}{2}\tfrac{3}{2}\Gamma(\tfrac{1}{2} + 1)} = \frac{\pi^{5/2}}{\tfrac{5}{2}\tfrac{3}{2}\tfrac{1}{2}\Gamma(\tfrac{1}{2})} = \frac{\pi^{5/2}}{\tfrac{5}{2}\tfrac{3}{2}\tfrac{1}{2}\sqrt{\pi}} = \frac{8}{15}\pi^2.$$

Figure 12.6 is the plot of the continuous form of figure 12.5, this time for n up to 20. It shows a little more clearly that the maximum occurs a little to the right of $n = 5$ and calculus should help us to find the coordinates of that point, provided that we can differentiate the components of

$$V_n(1) = \frac{\pi^{n/2}}{\Gamma(\tfrac{1}{2}n + 1)}$$

with respect to the continuous variable n.

The top of the fraction is easy to deal with using the fact that $a^b = e^{b\ln a}$ so that the formula becomes

$$V_n(1) = \frac{\pi^{n/2}}{\Gamma(\tfrac{1}{2}n + 1)} = \frac{e^{(n/2)\ln\pi}}{\Gamma(\tfrac{1}{2}n + 1)}.$$

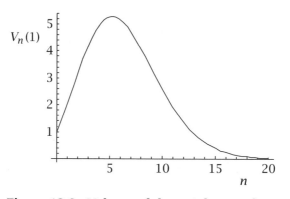

Figure 12.6. Volume of the unit hypersphere compared with continuous dimensions.

The bottom of the fraction requires us to differentiate the Gamma function. We have no need to look closely at the implications of this and we will merely write the derivative in the usual way as $\Gamma'(x)$; what we will need is the even more exotic Digamma function $\Psi(x)$, which is defined by

$$\Psi(x) = \frac{d}{dx} \ln \Gamma(x) = \frac{\Gamma'(x)}{\Gamma(x)}$$

and some powerful mathematical software to evaluate it.

Using the standard chain and quotient rules, the calculations are

$$\frac{dV_n(1)}{dn} = \frac{\Gamma(\frac{1}{2}n + 1)\frac{1}{2} \ln \pi e^{(n/2)\ln \pi} - e^{(n/2)\ln \pi} \frac{1}{2}\Gamma'(\frac{1}{2}n + 1)}{[\Gamma(\frac{1}{2}n + 1)]^2}$$

$$= \frac{\Gamma(\frac{1}{2}n + 1)\frac{1}{2} \ln \pi \pi^{n/2} - \pi^{n/2}\frac{1}{2}\Gamma'(\frac{1}{2}n + 1)}{[\Gamma(\frac{1}{2}n + 1)]^2}.$$

The requirement that $dV_n(1)/dn = 0$ means that

$$\Gamma(\tfrac{1}{2}n + 1)\tfrac{1}{2} \ln \pi \pi^{n/2} - \pi^{n/2}\tfrac{1}{2}\Gamma'(\tfrac{1}{2}n + 1) = 0$$

and so

$$\ln \pi - \frac{\Gamma'(\tfrac{1}{2}n + 1)}{\Gamma(\tfrac{1}{2}n + 1)} = 0 \quad \text{and} \quad \Psi(\tfrac{1}{2}n + 1) = \ln \pi.$$

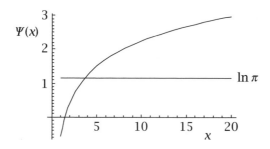

Figure 12.7. The Digamma function.

Figure 12.7 shows a plot of the Digamma function and the horizontal line at $\ln \pi$. To find $\frac{1}{2}n + 1$ and hence n we need computational help to establish that $n = 5.256\,946\,4\ldots$ and so is the dimension in which the unit hypersphere has a maximum volume and, substituting this value back into the formula, gives that maximum volume as $5.277\,768\ldots$.

So, a hypersphere of radius 1 achieves its maximum volume in $5.256\,946\,4\ldots$-dimensional space.

What must the radius of the sphere be to achieve its maximal volume in precisely five-dimensional space, or for that matter what must the radius of the sphere be to achieve its maximal volume in any other integral dimensional space? To answer these questions we need to consider the general $V_n(R)$ and differentiate it with respect to n just as we have done previously. The almost identical calculations are

$$V_n(R) = \frac{\pi^{n/2}}{\Gamma(\frac{1}{2}n + 1)} R^n = \frac{(\pi R^2)^{n/2}}{\Gamma(\frac{1}{2}n + 1)} = \frac{e^{(n/2)\ln(\pi R^2)}}{\Gamma(\frac{1}{2}n + 1)},$$

$$\frac{dV_n(R)}{dn} = \frac{\Gamma(\frac{1}{2}n + 1)\frac{1}{2}\ln(\pi R^2)e^{(n/2)\ln(\pi R^2)}}{[\Gamma(\frac{1}{2}n + 1)]^2}$$

$$- \frac{e^{(n/2)\ln(\pi R^2)}\frac{1}{2}\Gamma'(\frac{1}{2}n + 1)}{[\Gamma(\frac{1}{2}n + 1)]^2}$$

$$= \frac{\Gamma(\frac{1}{2}n + 1)\frac{1}{2}\ln(\pi R^2)(\pi R^2)^{n/2}}{[\Gamma(\frac{1}{2}n + 1)]^2}$$

$$- \frac{(\pi R^2)^{n/2}\frac{1}{2}\Gamma'(\frac{1}{2}n + 1)}{[\Gamma(\frac{1}{2}n + 1)]^2},$$

Table 12.2. Radius for maximal volume.

n	R
2	$0.696\,998\ldots$
3	$0.801\,888\ldots$
4	$0.894\,963\ldots$
5	$0.979\,428\ldots$
6	$1.057\,27\ldots$
7	$1.129\,83\ldots$
8	$1.198\,05\ldots$
9	$1.262\,61\ldots$
10	$1.324\,05\ldots$

and $dV_n(R)/dn = 0$ requires that

$$\Gamma(\tfrac{1}{2}n + 1)\tfrac{1}{2}\ln(\pi R^2)(\pi R^2)^{n/2} - (\pi R^2)^{n/2}\tfrac{1}{2}\Gamma'(\tfrac{1}{2}n + 1) = 0$$

and so

$$\Gamma(\tfrac{1}{2}n + 1)\ln(\pi R^2) - \Gamma'(\tfrac{1}{2}n + 1) = 0$$

and

$$\ln(\pi R^2) - \frac{\Gamma'(\tfrac{1}{2}n + 1)}{\Gamma(\tfrac{1}{2}n + 1)} = 0.$$

We have the general condition that

$$\Psi(\tfrac{1}{2}n + 1) = \ln(\pi R^2).$$

Table 12.2 shows the values of R for which a hypersphere of that radius achieves its maximal volume in low integral dimensional space. That is, a hypersphere of radius $0.696\,998\ldots$ achieves its maximum volume in two dimensions, one of radius $0.801\,888\ldots$ achieves its maximum volume in three dimensions, etc.

Sums of Volumes

The transcendental Gelfond Constant, e^π, appears naturally when we probe a little further into unit hypersphere volumes and investigate (ignoring units of measurement) the total volume of the infinite sequence of them. Since

$$V_n(1) = \frac{\pi^{n/2}}{\Gamma(\frac{1}{2}n + 1)} \to 0 \quad \text{as } n \to \infty,$$

there is at least a chance that there is a value to $\sum_{n=1}^{\infty} V_n(1)$ and, if we evaluate the finite sum for a large number of terms, our optimism seems to be well founded with that sum equal to 44.999 326 089 382 855 366 . . . .

To find a closed form for this we will again need to consider even and odd dimensions separately.

Recall that, for n even, we may write $V_n(1) = \pi^{n/2}/(\frac{1}{2}n)!$ for $n = 2, 4, 6, \ldots$ and, if $n = 2m$, we have that $V_{2m}(1) = \pi^m/m!$ for $m = 1, 2, 3, \ldots$. This means that

$$\sum_{n \text{ even}} V_n(1) = \sum_{m=1}^{\infty} V_{2m}(1) = \sum_{m=1}^{\infty} \frac{\pi^m}{m!} = e^\pi - 1$$

and we have the promised appearance of Gelfond's constant.

Matters are far more complicated if n is odd. Now we have that $V_n(1) = \pi^{n/2}/\Gamma(\frac{1}{2}n + 1)$ for $n = 1, 3, 5, \ldots$, and, if $n = 2m - 1$, we have $V_{2m-1}(1) = \pi^{m-1/2}/\Gamma(m + \frac{1}{2})$ for $m = 1, 2, 3, \ldots$ and the sum of the volumes is now

$$\sum_{n \text{ odd}} V_n(1) = \sum_{m=1}^{\infty} V_{2m-1}(1) = \sum_{m=1}^{\infty} \frac{\pi^{m-1/2}}{\Gamma(m + \frac{1}{2})},$$

which is altogether more challenging.

In fact, we can eliminate the Gamma function from the expression by using a result which connects it to another exotic function, the double factorial $N!!$, which is defined by

$$N!! = \begin{cases} N(N-2) \cdots 5 \times 3 \times 1, & N(>0) \text{ odd}, \\ N(N-2) \cdots 6 \times 4 \times 2, & N(>0) \text{ even}, \\ 1, & N = -1, 0. \end{cases}$$

Using the standard properties of the Gamma function, it is not too hard to show that $\Gamma(m + \frac{1}{2}) = ((2m - 1)!!/2^m)\sqrt{\pi}$ and this makes

$$\sum_{n \text{ odd}} V_n(1) = \sum_{m=1}^{\infty} \frac{\pi^{m-1/2}}{\Gamma(m + \frac{1}{2})}$$

$$= \sum_{m=1}^{\infty} \frac{\pi^{m-1/2}}{\{((2m - 1)!!/2^m)\sqrt{\pi}\}}$$

$$= \sum_{m=1}^{\infty} \frac{2^m \pi^{m-1}}{(2m - 1)!!}$$

and it is easier still to show that $(2m - 1)!! = (2m)!/(2^m m!)$, and this makes

$$\sum_{n \text{ odd}} V_n(1) = \sum_{m=1}^{\infty} \frac{2^m \pi^{m-1}}{(2m - 1)!!}$$

$$= \sum_{m=1}^{\infty} \frac{2^m \pi^{m-1}}{\{(2m)!/(2^m m!)\}}$$

$$= \sum_{m=1}^{\infty} \frac{2^{2m} \pi^{m-1} m!}{(2m)!}.$$

Now that we have the sum expressed in more elementary terms it is still far from obvious whether or not this series for odd n has a closed form, as the much simpler one did for even n. If we begin to write out the series explicitly, we have

$$2 + \frac{4}{3}\pi + \frac{8}{15}\pi^2 + \frac{16}{105}\pi^3 + \frac{52}{945}\pi^4 + \cdots$$

and those coefficients do not look particularly promising: a look in a standard mathematical handbook reveals nothing. In fact, the series does have a closed form, and to approach it we will engage in a common mathematical technique: the optimistic guess. Since e^π appears in the expression for even n, it might just do so here and if it does the most reasonable form of its appearance would be

$$\sum_{m=1}^{\infty} \frac{2^{2m} \pi^{m-1} m!}{(2m)!} = e^\pi S(\pi),$$

where $S(\pi)$ is an infinite series in π. To find the form that this series must have, we need to rewrite the expression and expand both sides to compare coefficients:

$$2 + \tfrac{4}{3}\pi + \tfrac{8}{15}\pi^2 + \tfrac{16}{105}\pi^3 + \tfrac{52}{945}\pi^4 + \cdots$$
$$= (1 + \pi + \tfrac{1}{2}\pi^2 + \tfrac{1}{6}\pi^3 + \tfrac{1}{24}\pi^4 + \cdots)$$
$$\times (a_0 + a_1\pi + a_2\pi^2 + a_3\pi^3 + a_4\pi^4 + \cdots),$$

which leads to the sequence of coefficients, $a_0 = 1$, $a_1 = -\tfrac{2}{3}$, $a_2 = \tfrac{1}{5}$, $a_3 = -\tfrac{1}{21}$, $a_4 = \tfrac{1}{108}, \ldots$, and our new series is

$$S(\pi) = 2 - \tfrac{2}{3}\pi + \tfrac{1}{5}\pi^2 - \tfrac{1}{21}\pi^3 + \tfrac{1}{108}\pi^4 + \cdots .$$

Even these coefficients promise little, but a second look in that mathematical handbook reveals the error function, $\mathrm{Erf}(x)$, with its series form

$$\mathrm{Erf}(x) = \frac{1}{\sqrt{\pi}}(2x - \tfrac{2}{3}x^3 + \tfrac{1}{5}x^5 - \tfrac{1}{21}x^7 + \tfrac{1}{108}x^9 - \cdots).$$

In fact, it is defined by

$$\mathrm{Erf}(x) = \frac{2}{\sqrt{\pi}} \int_0^x e^{-t^2}\, dt$$

and comes from the theory of the normal distribution in statistics. Using the series expansion of e^{-t^2} and integrating term by term results in the series form. Evaluate at $x = \sqrt{\pi}$ and we have precisely

$$\mathrm{Erf}(\sqrt{\pi})$$
$$= \frac{1}{\sqrt{\pi}}(2\sqrt{\pi} - \tfrac{2}{3}\sqrt{\pi}^3 + \tfrac{1}{5}\sqrt{\pi}^5 - \tfrac{1}{21}\sqrt{\pi}^7 + \tfrac{1}{108}\sqrt{\pi}^9 - \cdots)$$
$$= (2 - \tfrac{2}{3}\pi + \tfrac{1}{5}\pi^2 - \tfrac{1}{21}\pi^3 + \tfrac{1}{108}\pi^4 - \cdots).$$

Our $S(\pi)$ is $\mathrm{Erf}(\sqrt{\pi})$ and therefore we have

$$\sum_{n \text{ odd}} V_n(1) = \sum_{m=1}^{\infty} \frac{2^{2m}\pi^{m-1}m!}{(2m)!} = e^{\pi}\, \mathrm{Erf}(\sqrt{\pi}).$$

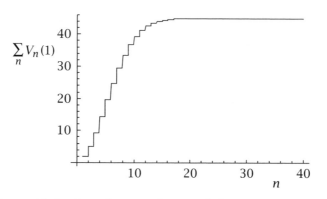

Figure 12.8. Cumulative volumes of the unit hypercubes.

And this makes

$$\sum_n V_n(1) = \begin{cases} e^{\pi} - 1, & n \text{ even,} \\ e^{\pi} \operatorname{Erf}(\sqrt{\pi}), & n \text{ odd.} \end{cases}$$

And the total volume

$$\sum_n V_n(1) = (e^{\pi} - 1) + e^{\pi} \operatorname{Erf}(\sqrt{\pi}) = e^{\pi}(1 + \operatorname{Erf}(\sqrt{\pi})) - 1.$$

Of course, we have not rigorously proved this but from what we have seen it is at least feasible and evaluation of this exact expression to 44.999 326 089 382 855 366... can only add to our confidence. It is the case that not too much extra analysis would prove it to be so.

In fact, we can see from figure 12.8, which shows a continuous plot of the cumulative sum, that the convergence is all but accomplished by the twentieth term.

Surface Area in Hyperdimensions

The surface area $A_n(R)$ of the hypersphere whose volume is

$$V_n(R) = \frac{\pi^{n/2}}{\Gamma(\frac{1}{2}n + 1)} R^n$$

is simply the derivative of the expression with respect to R, so

$$A_n(R) = \frac{n\pi^{n/2}}{\Gamma(\frac{1}{2}n + 1)} R^{n-1} = \frac{n\pi^{n/2}}{\frac{1}{2}n\Gamma(\frac{1}{2}n)} R^{n-1} = \frac{2\pi^{n/2}}{\Gamma(\frac{1}{2}n)} R^{n-1}.$$

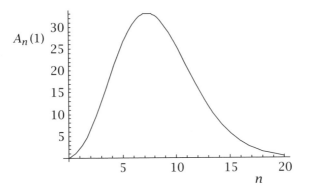

Figure 12.9. Surface area of the unit hypersphere
compared with continuous dimensions.

Table 12.3. Surface areas of hyperspheres.

n	$A_n(R)$	$A_n(1)$
2	$2\pi R$	$2\pi = 6.28318\ldots$
3	$4\pi R^2$	$4\pi = 12.5663\ldots$
4	$2\pi^2 R^3$	$2\pi^2 = 19.7392\ldots$
5	$\frac{8}{3}\pi^2 R^4$	$\frac{8}{3}\pi^2 = 26.31894\ldots$
6	$\pi^3 R^5$	$\pi^3 = 31.00627\ldots$
7	$\frac{16}{15}\pi^3 R^6$	$\frac{16}{15}\pi^3 = 33.07336\ldots$
8	$\frac{1}{3}\pi^4 R^7$	$\frac{1}{3}\pi^4 = 32.4696\ldots$

In particular,

$$A_n(1) = \frac{2\pi^{n/2}}{\Gamma(\frac{1}{2}n)}.$$

Table 12.3 lists the first few values of the hypersurface areas
and indicates a peak at $n = 7$ for $A_n(1)$. Figure 12.9 shows a plot
of $A_n(1)$ against a continuous n and indicates that the surface
area of the hypersphere does peak at around $n = 7$ and that it
also tends to 0 as n increases. Entirely similar calculations to
the ones previously made show that for maximal $A_n(1)$ n satisfies $\Psi(\frac{1}{2}n) = \ln\pi$, which tells us that the maximum is actually
achieved when $n = 7.25695\ldots$ and takes the value $33.1612\ldots$.

Table 12.4. Radius for maximal surface area.

n	R
2	0.422 751 ...
3	0.574 578 ...
4	0.696 998 ...
5	0.801 888 ...
6	0.894 963 ...
7	0.979 428 ...
8	1.057 27 ...
9	1.129 83 ...
10	1.198 05 ...

Almost repeating the argument for volume, if we wish to calculate the radius R of the hypersphere which has maximal surface area in each integral dimension, we use calculus on $A_n(R)$ and this gives rise to the equation

$$\Psi(\tfrac{1}{2}n) = \ln(\pi R^2),$$

which generates table 12.4.

In summary, the unit hypersphere has a maximum volume of 5.277 768 ... in 5.256 946 4 ...-dimensional space and a maximum surface area of 33.1612 ... in 7.256 95 ...-dimensional space. Further, a hypersphere of radius 0.696 998 ... achieves its maximum volume, and one of radius 0.422 751 ... its maximum surface area in two dimensions; one of radius 0.801 888 ... achieves its maximum volume, and one of radius 0.574 578 ... its maximum surface area in three dimensions, etc.

(Using very similar techniques as before, the sum of the surface areas of hyperspheres can be shown to be $2\sqrt{2\pi}\,\mathrm{e}^\pi$ for even dimensions and $2(1 + \pi \mathrm{e}^\pi \mathrm{Erf}(\sqrt{\pi}))$ for odd dimensions, making the total

$$2\sqrt{2\pi}\,\mathrm{e}^\pi + 2(1 + \pi \mathrm{e}^\pi \mathrm{Erf}(\sqrt{\pi}))$$
$$= 261.635\,258\,772\,474\,984\,53 \ldots .$$

A similar plot to figure 12.8 again shows that the convergence is all but accomplished by the twentieth term.)

Table 12.5. Distribution of volume in a hypersphere.

n	Volume
2	36
3	49
4	59
5	67
10	89
20	99
30	100

The Distribution of Volume

Is any of this useful? A good answer is, who cares? That said, there are implications of some of the strange behaviour of hyperspace to the theory of sampling in large numbers of variables, and the many mathematical ideas which depend on the techniques. We will not discuss them here but we will show an area which causes problems.

The volume of the n-dimensional hypersphere with radius R is, of course,

$$V_n(R) = \frac{\pi^{n/2}}{\Gamma(\frac{1}{2}n + 1)} R^n.$$

Now we ask the question, Where is this volume? To answer this, we will initially be particular and ask the question, How much of the volume of the hypersphere is at a distance of 20% from its surface?

The answer, as a percentage to the nearest whole number and for varying dimensions, is given in table 12.5 and the figures clearly show that the volume near the surface is fast approaching 100%.

In general, the amount of volume near the surface of the hypersphere of radius R can be measured by the difference between the volume of the hypersphere and the volume of the hypersphere of radius $R(1 - \varepsilon)$, where ε is taken to be small (in table 12.5 $\varepsilon = 0.2$). Compare this quantity with the volume of the hypersphere itself and we have the fraction

$$\frac{V_n(R) - V_n(R(1 - \varepsilon))}{V_n(R)}$$

$$= \left(\frac{\pi^{n/2}}{\Gamma(\frac{1}{2}n + 1)} R^n - \frac{\pi^{n/2}}{\Gamma(\frac{1}{2}n + 1)} [R(1 - \varepsilon)]^n \right) \bigg/ \frac{\pi^{n/2}}{\Gamma(\frac{1}{2}n + 1)} R^n$$

$$= 1 - (1 - \varepsilon)^n \xrightarrow[n \to \infty]{} 1.$$

That the asymptotic limit is 1 shows that, whatever volume there is in a high-dimensional hypersphere (and there isn't much), it is ever more concentrated at its surface. Also, since, for all R,

$$V_n(R) \xrightarrow[n \to \infty]{} 0,$$

inscribe the hypersphere in the n-dimensional hypercube of side $2R$ and we see that most of the hypercube's volume is concentrated at its corners.

Our discussion has concentrated on a few particular areas of the counterintuitive nature of hyperdimensions and we could mention many other manifestations and implications: the full story is big enough to fill books (and has done so). To pursue the matter further in one important direction, the reader is encouraged to research the term 'the *curse of dimensionality*', which was coined by the American mathematician Richard Bellman in 1961.

Chapter 13

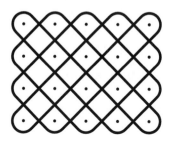

FRIDAY THE 13TH

I'm a great believer in luck and I find the harder I work, the more I have of it.

Thomas Jefferson

A Letter to the *Times*

The bottom right-hand slot of the letters page of the London *Times* is often reserved for offbeat or amusing correspondence and was occupied on Friday, 13 February 1970, by the following:

Sir,

If, as some of your recent correspondents suggest, eccentricity is one of the criteria for publication of letters to the *Times*, you may be willing to allow me, on this doubly unlucky date of Friday the Thirteenth of February, to remind any superstitious among your readers that the 13th day of the month falls more frequently on a Friday than upon any other day of the week.

This at first sight unbelievable property of the calendar as set up is exemplified by the present year, when no fewer than three of the 13th days of the month fall on a Friday; namely in February, March and November – 25 per cent; whereas the average figure would be only 14 per cent.

In case this excess causes alarm and despondency, it may be some consolation that the balance is slightly redressed by the fact that the first day of a new century can never fall upon a Friday. And incidentally, but for different reasons, neither can Ascension Day nor Pancake Day!

Yours truly,

Raymond A. Lyttleton, St. John's College, Cambridge.

This chapter is a study of the letter and matters related to it.

Superstition

Concern about, or even fear of, the number 13 has been given the seemingly unpronounceable name *triskaidekaphobia*, a Greek compound made from the following parts: *tris*, 'three'; *kai*, 'and'; *deka*, 'ten' (which makes thirteen); plus *phobia*, 'fear'. It appears to date from 1911, when it appeared in I. H. Coriat's *Abnormal Psychology*.

There is any number of justifications for 13 being unlucky: there were 13 present at the Last Supper, in Norse mythology there were 13 present at a banquet in Valhalla when Balder (son of Odin) was slain, which led to the downfall of the gods; Hesiod wrote in *Works and Days* that the thirteenth day is unlucky for sowing, but favourable for planting. What is assuredly not apocryphal is the near-catastrophic explosion on the Moon rocket *Apollo 13*, which occurred later in the same year as Lyttleton's letter, on 13 April 1970 (a Monday), two days after its launch from the Kennedy Space Center at 14:13:00 EST (13:13:00 CST) from launch pad 39 (3×13).

Naturally, to balance matters we have triskaidekamania, which is an excessive enthusiasm for the number 13. And, to shift the blame elsewhere, tetraphobia, a fear of the number 4, which is most common in East Asian countries (since the pronunciation of the number is close to that of a word for death).

Yet matters are worse, since the letter also refers to Friday and so touches on the condition of paraskevidekatriaphobia, or fear of Friday the 13th. Again, the word is made more natural by examining its Greek root, *paraskevi*, 'Friday', and *dekatria*, 'thirteen', with, of course, *phobia* at the end.

The origins of Friday superstitions are many: Eve tempted Adam to eat the forbidden fruit, the Great Flood occurred, the start of the linguistic confusion at the Tower of Babel, the destruction of Solomon's Temple and the death of Jesus Christ all, by tradition, took place on a Friday.

There is a story that, in the eighteenth century, the British government tried to relieve its fiercely superstitious sailors of their fear of sailing on Fridays by building a ship called the HMS *Friday*. However, when she set sail one Friday morning on the 13th, under Captain Jim Friday's command, she was never seen again.

Whatever the probity of the reasons, our Western civilization is stuck with the belief that both Friday and the number 13 are unlucky and in combination doubly so; such a hold does this belief have that skyscrapers are commonly known to have floor numbering that goes from 12 to 14, and ships still delay sailing to avoid Fridays. Franklin Delano Roosevelt is said to have suffered from triskaidekaphobia but, judging from the final line of Samuel Pepys's diary entry for Friday, 13 July 1660, he eschewed triskaidekaphobia and indeed paraskevidekatriaphobia:

> To bed with the greatest quiet of mind that I have had a great while.

Now we will look at the reason that

> the 13th day of the month falls more frequently on a Friday than upon any other day of the week.

The mathematics involved is contained in two of the more strange of the elementary formulae to come from the minds of mathematicians, the first from the greatest mind of them all.

Table 13.1. Values of e for each month m.

m	1	2	3	4	5	6	7	8	9	10	11	12
e	0	3	2	5	0	3	5	1	4	6	2	4

Table 13.2. Values of f for each century c.

$c \bmod 4$	c	f
0	16, 20, ...	0
1	17, 21, ...	5
2	18, 22, ...	3
3	19, 23, ...	1

Gauss's Formula

This opaque formula is one of several that have been developed to establish the day of the week of a given date and, for the reader who would like to see exactly where it comes from, it has been analysed, in particular, by Berndt Schwerdtfeger (in his Internet article *Gauss calandar formula for the day of the week*[1]): we will simply state it and put it to use, but to state the formula we need to establish some notation.

The variable w enumerates the day of the week, with $w = 1$ corresponding to Monday, $w = 2$ corresponding to Tuesday, etc. The variable d is the day of the month, so $d \in \{1, 2, 3, \ldots, 31\}$. The variable m is the number of the month, beginning with January and finishing in December, so $m \in \{1, 2, 3, \ldots, 12\}$. The variable y is the year, given as a four-digit integer, $c = \lfloor y/100 \rfloor$ is the two-digit century and $g = y - 100c \in \{0, 1, 2, \ldots, 99\}$ is the two-digit year of the century.

The month m has associated with it a variable e, the values of which are given in table 13.1. The century c has associated with it a variable f, the values of which are given in table 13.2.

Lastly, there is one more rule: if $m = 1$ or 2, then y is replaced by $y - 1$ in the calculations of c and g.

---

[1]http://berndt-schwerdtfeger.de/articles.html

With all of this mathematical alchemy in place, Gauss's formula for the day of the week of any date in the Gregorian calendar is

$$w = d + e + f + g + \lfloor \tfrac{1}{4}g \rfloor \quad \text{mod } 7.$$

In all of this, $\lfloor x \rfloor$ is the floor function, defined as the greatest integer less than or equal to x, as we have seen earlier in the book.

The Big Count

To establish the fact that the 13th of the month is more likely to fall on a Friday than on any other day of the week, we need to look carefully at the implications of the adoption of the Gregorian calendar. The calendar is named after Pope Gregory XIII, who instituted it in 1582 when he decreed that the day after 4 October 1582 would be 15 October 1582 (to considerable public consternation). It is a modified version of the Julian calendar, named after Julius Caesar, and which, by that time, was badly out of synchronization. In the Gregorian calendar the leap year that occurs once every four years is omitted in years divisible by 100 but not divisible by 400. This means that, for example, 2000 was a leap year (since it is divisible by 400) but 2100 will not be (since the number is divisible by 100 but not 400). This carries the important consequence that the Gregorian calendar repeats itself precisely every 400 years, since the number of days in 400 Gregorian years is $100(3 \times 365 + 366) - 3 = 146\,097$ and this is an exact number of weeks, since $146\,097$ is exactly divisible by 7.

To compile our data, therefore, we simply need to count up the frequencies with which the 13th of the month falls on each day of the week in a cycle of 400 years, and to do this we can program a computer to use Gauss's day of the week formula; the results of this computer calculation are given in table 13.3.

A look at the bottom row reveals that the 13th of the month falls on a Friday 688 times out of the possible 4800, just beating Wednesday and Sunday. Professor Lyttleton is vindicated, as he is with the remainder of his statements.

Table 13.3. Frequencies with which the 13th of the month falls on each day of the week in a cycle of 400 years.

	Mon.	Tue.	Wed.	Thu.	Fri.	Sat.	Sun.	Total
January	57	57	58	56	58	56	58	400
February	58	56	58	57	57	58	56	400
March	56	58	57	57	58	56	58	400
April	58	56	58	56	58	57	57	400
May	57	57	58	56	58	56	58	400
June	58	56	58	57	57	58	56	400
July	58	56	58	56	58	57	57	400
August	58	57	57	58	56	58	56	400
September	56	58	56	58	57	57	58	400
October	57	58	56	58	56	58	57	400
November	56	58	57	57	58	56	58	400
December	56	58	56	58	57	57	58	400
Total	685	685	687	684	**688**	684	687	4800

The Remainder of the Letter

Lyttleton's final paragraph contains the comforting observation, 'that the first day of a new century can never fall upon a Friday'. Care must be taken with what he means by the statement; for example, 1 January 2100 will be a Friday. He has taken the view that, since there is no year 0, the first day of the new century will have the year part of 01, for example, 1 January 2101. In fact, more can be said, in that *the first day of a new century can never fall on a Friday, Wednesday or Sunday.*

To see this we should agree that, since we are interested in January, $m = 1$ and so we must use a year part of 00 for the calculation of c and g, making $g = 0$. The calculation is, then, $w = (1 + 0 + f + 0 + 0) \bmod 7 = (1 + f) \bmod 7$. If we in turn substitute $\{0, 1, 3, 5\}$ (the four possible values of f), we arrive at the equations

$$w = 1 \bmod 7, \quad w = 2 \bmod 7, \quad w = 4 \bmod 7, \quad w = 6 \bmod 7,$$

the solutions for which are $w = 1, 2, 4, 6$, respectively. This

Table 13.4. When in the year the 13th of the month occurs.

Month	Non-leap year Day of year	mod 7	Leap year Day of year	mod 7
January	13	6	13	6
February	44	2	44	2
March	72	2	73	3
April	103	5	104	6
May	133	0	134	1
June	164	3	165	4
July	194	5	195	6
August	225	1	226	2
September	256	4	257	5
October	286	6	287	0
November	317	2	318	3
December	347	4	348	5

means that 1 January of a new century can only start on a Monday, Tuesday, Thursday or Saturday; omitted from the list is that dreaded Friday and Wednesday and Sunday as well.

In Lyttleton's second paragraph, he refers to the frequency with which Friday the 13th occurs in a year and we will look more closely at this aspect of that most unpropitious day.

If we take a non-leap year and number the days from 1 January to 31 December from 1 to 365, we can tabulate which day of the year corresponds to the 13th of each month. For example, 13 January is the 13th day of the year, 13 February is the 44th day of the year, 13 March is the 72nd day of the year, etc. If we reduce these numbers modulo 7, we can ascertain on which day of the week each date of the 13th occurred, provided we know on what day of the week 1 January occurs. The same enumeration can be used for leap years and table 13.4 summarizes the results.

Now we need to consider the implications for Fridays, given that 1 January falls on each of the days of the week, to arrive at the first two columns of table 13.5. Cross-referencing with table 13.4 results in the third column of table 13.5, which lists

Table 13.5. Restrictions for a non-leap year.

If 1 January is on ...	Then Fridays will reduce modulo 7 to ...	So, Friday the 13th will occur in the months of ...
Sunday	6	January, October
Monday	5	April, July
Tuesday	4	September, December
Wednesday	3	June
Thursday	2	February, March, November
Friday	1	August
Saturday	0	May

Table 13.6. Restrictions for a leap year.

If 1 January is on ...	Then Fridays will reduce modulo 7 to ...	So, Friday the 13th will occur in the months of ...
Sunday	6	January, April, July
Monday	5	September, December
Tuesday	4	June
Wednesday	3	March, November
Thursday	2	February, August
Friday	1	May
Saturday	0	October

the months having Friday the 13th in them. Table 13.6 again repeats the whole thing for leap years.

From these last two tables we can deduce the following:

- There is at least one Friday the 13th in every year.
- The greatest number of occurrences of Friday the 13th in any year is three. Also, these must occur in February, March and November in a non-leap year and January, April and July in a leap year.

The year 1970 was not a leap year and so the triplet of months had to be as Lyttleton had it. Since every month appears in the right-hand columns of tables 13.5 and 13.6, the average number of Friday the 13ths in every year is simply $\frac{1}{7}$, or 14%; the full calculation for a non-leap year is

$$(\tfrac{1}{7} \times \tfrac{2}{12}) + (\tfrac{1}{7} \times \tfrac{2}{12}) + (\tfrac{1}{7} \times \tfrac{2}{12}) + (\tfrac{1}{7} \times \tfrac{1}{12}) + (\tfrac{1}{7} \times \tfrac{3}{12}) + (\tfrac{1}{7} \times \tfrac{1}{12}) + (\tfrac{1}{7} \times \tfrac{1}{12})$$
$$= \tfrac{1}{7} \times \tfrac{12}{12} = \tfrac{1}{7}.$$

- The only possibility for consecutive months having Friday the 13th in them is February and March and this can only occur in a non-leap year. The next few years are 2009, 2015 and 2026.

The final part of the letter is whimsical, but it does provide the motivation for the second recondite algorithm, one which computes the day on which Easter Sunday falls in any particular year. With Easter Sunday defined as the first Sunday after the first full Moon to occur after the vernal equinox (21 March), it is small surprise that the computation of its date is somewhat complex; it is no surprise at all that Gauss provided an algorithm for this computation too. That said, his procedure is a little inelegant in that there are exceptions to consider, and we choose to present an algorithm which has none such and which seems to have its origins in the early nineteenth century with the French mathematical astronomer and contemporary of Gauss, Jean Baptiste Delambre; it is valid for all Gregorian years, that is, for 1583 and beyond.

The left-hand column of table 13.7 provides the algorithm and the right-hand column a single check of it. (The months are numbered naturally from January (1) to December (12).)

The calculation tell us that Easter Sunday in 2005 fell on 27 March, and it did! In fact, it can be shown that Easter Sunday can fall on one of 35 dates: 22 March to 25 April.

And what of Ascension Day? This is the 40th day after Easter Sunday and commemorates the Ascension of Christ into heaven, according to Mark 16:19, Luke 24:51 and Acts 1:2. And, to save counting, an alternative name provides a strong hint that it cannot fall on a Friday: Holy Thursday (not to be confused with Maundy Thursday).

And Pancake Day? This is the English term for the day before the beginning of Lent, 47 days before Easter Sunday: no counting

Table 13.7. A test of the Delambre formula.

Algorithm	$Y = 2005$
$a = Y \bmod 19$	$a = 2005 \bmod 19 = 10$
$b = \left\lfloor \dfrac{Y}{100} \right\rfloor$	$b = \left\lfloor \dfrac{2005}{100} \right\rfloor = 20$
$c = Y \bmod 100$	$c = 2005 \bmod 100 = 5$
$d = \left\lfloor \dfrac{b}{4} \right\rfloor$	$d = \left\lfloor \dfrac{20}{4} \right\rfloor = 5$
$e = b \bmod 4$	$e = 20 \bmod 4 = 0$
$f = \left\lfloor \dfrac{b + 8}{25} \right\rfloor$	$f = \left\lfloor \dfrac{20 + 8}{25} \right\rfloor = 1$
$g = \left\lfloor \dfrac{b - f + 1}{3} \right\rfloor$	$g = \left\lfloor \dfrac{20 - 1 + 1}{3} \right\rfloor = 6$
$h = (19a + b - d \\ \quad - g + 15) \bmod 30$	$h = (19 \times 10 + 20 \\ \quad - 5 - 6 + 15) \bmod 30 \\ \quad = 4$
$i = \left\lfloor \dfrac{c}{4} \right\rfloor$	$i = \left\lfloor \dfrac{5}{4} \right\rfloor = 1$
$k = c \bmod 4$	$k = 5 \bmod 4 = 1$
$l = (32 + 2e + 2i \\ \quad - h - k) \bmod 7$	$l = (32 + 2 \times 0 + 2 \times 1 \\ \quad - 4 - 1) \bmod 7 = 1$
$m = \left\lfloor \dfrac{a + 11h + 22l}{451} \right\rfloor$	$m = \left\lfloor \dfrac{10 + 11 \times 4 + 22 \times 1}{451} \right\rfloor \\ \quad = 0$
$\text{Month} = \left\lfloor \dfrac{h + l - 7m + 114}{31} \right\rfloor$	$\text{Month} = \left\lfloor \dfrac{4 + 1 - 7 \times 0 + 114}{31} \right\rfloor \\ \quad = 3$
$\text{Day} = (h + l - 7m \\ \quad + 114) \bmod 31 + 1$	$\text{Day} = (4 + 1 - 7 \times 0 \\ \quad + 114) \bmod 31 + 1 \\ \quad = 27$

is needed here either; Lent begins on Ash Wednesday, one day after Shrove Tuesday, otherwise called Pancake Day.

And what of superstition? Eggs laid on Ascension Day are said never to go bad and will guarantee good luck for a household if

Table 13.8. The reuse of a calendar.

Years after a leap year	Reuse calendar in year(s)
0	$X + 28$
1	$X + 6$, $X + 17$, $X + 28$
2	$X + 11$, $X + 17$, $X + 28$
3	$X + 11$, $X + 22$, $X + 28$

placed in the roof, and in some parts of England the first pancake made on Pancake Day is given to chickens, to ensure their fertility during the year.

To end with we list, for the frugally minded, the years between 1901 and 2099 in which an old calendar can be reused, under the assumption that the interest is in which dates fall on which days of the week, rather than the irregular holidays such as Easter. Table 13.8 summarizes the information for year X, which is characterized by the number of years X is after a leap year.

Notice that the expression $X + 28$ occurs in all four rows, which means that a calendar can always be used every 28 years.

Incorporating the requirement that Easter will also be on the same day brings with it considerable serendipity. For example, the calendars for the years 1981 and 1987 are identical, including the date for Easter, whereas the calendar from 1940 will not be reusable until 5280!

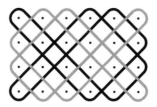

FRACTRAN

Everything should be made as simple as possible, but not simpler.

Albert Einstein

Mysterious Arithmetic

In chapter 6 we looked at one idea from the fertile and original mind of John Conway, and now we will look at a second, which, in typical whimsical style, he called 'fourteen fantastic fractions' in his joint publication with Richard Guy, *The Book of Numbers*. The idea appeared earlier, in his article *Fractran: A Simple Universal Programming Language for Arithmetic*, which constituted chapter 2 of the 1987 book *Open Problems in Communication and Computation* (ed. T. M. Cover and B. Gopinath), Springer, pp. 4-26. Further articles about the construction abound; we have used one of Richard Guy's from *Mathematics Magazine*: 'Conway's prime producing machine' (1983) 56:26–33.

The fractions in question are the seemingly arbitrarily ordered collection

$$\left\{ \begin{array}{cccccccccccccc} \frac{17}{91} & \frac{78}{85} & \frac{19}{51} & \frac{23}{38} & \frac{29}{33} & \frac{77}{29} & \frac{95}{23} & \frac{77}{19} & \frac{1}{17} & \frac{11}{13} & \frac{13}{11} & \frac{15}{14} & \frac{15}{2} & \frac{55}{1} \\ A & B & C & D & E & F & G & H & I & J & K & L & M & N \end{array} \right\}$$

with each labelled with a letter of the alphabet for easy reference.

We will now play a seemingly arbitrary game with the seemingly arbitrary fractions. Start with the integer 2 and multiply it
in turn, starting at A, by each of the fractions until we arrive at
a new integer: clearly, it is the fraction M which results in success, yielding 15 as the product. The process is repeated with
the new integer (15), again starting at A and continuing to the
first fraction to yield an integer product once again (N in this
case, yielding 825). This is repeated indefinitely, but each time an
exact power of 2 is reached, that power is noted. At this stage it
is not at all clear that a power of 2 will be reached at all, let alone
more than one of them; in fact, an infinite number of them will
be reached, and those powers form a very important sequence of
positive integers. With the intrigue thus built up, we will look at
the first stages of what results from this opaque process, listing
as a pair the current integer and the fraction which yields the
first integer product:

$$(2, M), \quad (15, N), \quad (825, E), \quad (725, F), \quad (1925, K),$$
$$(2275, A), \quad (425, B), \quad (390, J), \quad (330, E), \quad (290, F),$$
$$(770, K), \quad (910, A), \quad (170, B), \quad (156, J), \quad (132, E),$$
$$(116, F), \quad (308, K), \quad (364, A), \quad (68, I), \quad (4, M),$$

and we have reached the first power of 2 after 19 steps; that
power is itself 2. The reader can easily check the list with a cal-
culator but to proceed further really needs an appropriate computer program and using one confirms that the next power of 2
to be reached is 8 (after 69 steps) and the one after that is 32
(after 281 steps). For those interested, the full list of generated
integers up to this stage is given in figure 14.1.

A Mystery revealed

What is so special about the process? If we perform it repeatedly,
we arrive at the list of generated powers of 2:

$$4, \quad 8, \quad 32, \quad 128, \quad 2048, \quad 8192, \quad 131\,072, \ldots,$$

or, written in exponential form, $2^2, 2^3, 2^5, 2^7, 2^{11}, 2^{13}, 2^{17}, \ldots$ and
the exponents are none other than the prime numbers in order;

2, 15, 825, 725, 1925, 2275, 425, 390, 330, 290, 770,
910, 170, 156, 132, 116, 308, 364, 68, **4**, 30, 225, 12 375,
10 875, 28 875, 25 375, 67 375, 79 625, 14 875, 13 650,
2550, 2340, 1980, 1740, 4620, 4060, 10 780, 12 740, 2380,
2184, 408, 152, 92, 380, 230, 950, 575, 2375, 9625,
11 375, 2125, 1950, 1650, 1450, 3850, 4550, 850, 780,
660, 580, 1540, 1820, 340, 312, 264, 232, 616, 728, 136,
8, 60, 450, 3375, 185 625, 163 125, 433 125, 380 625,
1 010 625, 888 125, 2 358 125, 2 786 875, 520 625, 477 750,
89 250, 81 900, 15 300, 14 040, 11 880, 10 440, 27 720,
24 360, 64 680, 56 840, 150 920, 178 360, 33 320, 30 576,
5712, 2128, 1288, 5320, 3220, 13 300, 8050, 33 250,
20 125, 83 125, 336 875, 398 125, 74 375, 68 250, 12 750,
11 700, 9900, 8700, 23 100, 20 300, 53 900, 63 700, 11 900,
10 920, 2040, 1872, 1584, 1392, 3696, 3248, 8624, 10 192,
1904, 112, 840, 6300, 47 250, 354 375, 50 625, 2 784 375,
2 446 875, 6 496 875, 5 709 375, 15 159 375, 13 321 875,
35 371 875, 31 084 375, 82 534 375, 97 540 625, 18 221 875,
16 721 250, 3 123 750, 2 866 500, 535 500, 491 400, 91 800,
84 240, 71 280, 62 640, 166 320, 146 160, 388 080, 341 040,
905 520, 795 760, 2 112 880, 2 497 040, 466 480, 428 064,
79 968, 29 792, 18 032, 74 480, 45 080, 186 200, 112 700,
465 500, 281 750, 1 163 750, 704 375, 2 909 375, 11 790 625,
13 934 375, 2 603 125, 2 388 750, 446 250, 409 500, 76 500,
70 200, 59 400, 52 200, 138 600, 121 800, 323 400, 284 200,
754 600, 891 800, 166 600, 152 880, 28 560, 26 208, 4896,
1824, 1104, 4560, 2760, 11 400, 6900, 28 500, 17 250,
71 250, 43 125, 178 125, 721 875, 634 375, 1 684 375,
1 990 625, 371 875, 341 250, 63 750, 58 500, 49 500, 43 500,
115 500, 101 500, 269 500, 318 500, 59 500, 54 600, 10 200,
9360, 7920, 6960, 18 480, 16 240, 43 120, 50 960, 9520,
8736, 1632, 608, 368, 1520, 920, 3800, 2300, 9500, 5750,
23 750, 14 375, 59 375, 240 625, 284 375, 53 125, 48 750,
41 250, 36 250, 96 250, 113 750, 21 250, 19 500, 16 500,
14 500, 38 500, 45 500, 8500, 7800, 6600, 5800, 15 400,
18 200, 3400, 3120, 2640, 2320, 6160, 7280, 1360, 1248,
1056, 928, 2464, 2912, 544, **32**

Figure 14.1. The sequence to reach 32.

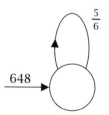

Figure 14.2. The loop corresponding to $\{\frac{5}{6}\}$.

in fact, incredible though it may seem, this process is nothing other than a prime-producing procedure: every prime will be generated, in order. Conway called this process PRIMEGAME.

The generation of prime numbers is a perfectly straightforward programming problem and therefore, in principle, a simple computational matter, but why should this arithmetic trick achieve what a programming language naturally succeeds in? The answer is that it really is a programming language in disguise.

To begin an explanation, we will frame the process in a more general context:

(1) Decide on an ordered list of fractions and a starting integer, N.

(2) Multiply the current integer (initially N) by the first fraction in the list for which the product is itself an integer and so obtain a new integer.

(3) Repeat step 2 until no product produces an integer, in which case the processes stops, or continue it indefinitely.

For example, suppose that our list is populated by a single fraction $\{\frac{5}{6}\}$ and suppose also that our starting integer is $N = 648$. We can represent this recycling process by the loop as shown in figure 14.2 and by the more formal pseudo-programming statement:

$$\text{line 1}: \quad \tfrac{5}{6} \rightarrow 1,$$

which we will take to mean 'multiply the input by $\frac{5}{6}$ and continue doing so for as long as the input is an integer by starting again at line 1'.

This seemingly random pair of choices of fraction and input causes the loop to be traversed exactly three times before fractions have to be introduced, which brings the process to a halt with $N = 375 = (\frac{5}{6})^3 \times 648$. So what?

In fact, $\frac{5}{6}$ and $N = 648$ are not at all random choices; it is no small matter that $\frac{5}{6} = \frac{5}{2 \times 3}$ and that $N = 648 = 2^3 \times 3^4$ is an example of a number which is of the form $N = 2^n \times 3^m$ for some nonnegative integers m and n. The factorization of $\frac{5}{6}$ shows that each multiplication by it will decrease the powers of 2 and 3 in the representation of N each by 1 and increase the power of 5 by 1, and this will continue for as long as the product is an integer; that final integer is $375 = 3 \times 5^3$.

If we consider the values of the powers of 2, 3 and 5 in the representation of N to be values held in the dynamic registers r_2, r_3 and r_5, respectively, in the general case we start with $r_2 = n$, $r_3 = m$, $r_5 = 0$ and finish with either $r_2 = 0$ or $r_3 = 0$ and $r_5 = \min(m, n)$; the contents of the 5 register finishes with the minimum of the two integers m and n and the process is seen to be precisely one of finding the minimum of two nonnegative integers. In terms of a typical, real programming language this is equivalent to

$r_2 := n;\quad r_3 := m;\quad r_5 := 0;$
While ($r_2 > 0$ and $r_3 > 0$) do
    Begin
      $r_2 := r_2 - 1;\quad r_3 := r_3 - 1;\quad r_5 := r_5 + 1;$
    End;
Print r_5; 'is the minimum of' m and n;

We can see that this simple process is therefore equivalent to a conventional programming algorithm.

Change the fraction to $\frac{10}{3} = \frac{2 \times 5}{3}$ for the same input of $N = 2^n \times 3^m$ as in figure 14.3 and the process adds 1 to each of r_2 and r_5 and subtracts 1 from r_3, until $r_3 = 0$, at which point $r_2 = m + n$ and $r_5 = m$ and the 2 register contains the sum of m and n; we have an adding machine. This time the equivalent code is

$r_2 := n;\quad r_3 := m;\quad r_5 := 0;$

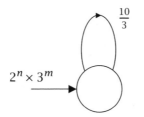

Figure 14.3. The loop corresponding to $\{\frac{10}{3}\}$.

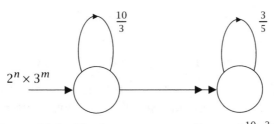

Figure 14.4. The loop corresponding to $\{\frac{10}{3}, \frac{3}{5}\}$.

While $r_3 > 0$ do
   Begin
      $r_2 := r_2 + 1$; $r_3 := r_3 - 1$; $r_5 := r_5 + 1$;
   End;
Print r_2; 'is the sum of' m and n;

We can tidy up this last solution and at the same time prepare it for generalization by introducing the second fraction of $\frac{3}{5}$ at its end and so form the fraction list $\{\frac{10}{3}, \frac{3}{5}\}$. Figure 14.4 represents this, with the agreed convention that single arrow routes have precedence over double arrow routes.

This time the pseudo-code is

$$\text{line 1}: \quad \frac{10}{3} \to 1, \quad \frac{1}{1} \to 2,$$
$$\text{line 2}: \quad \frac{3}{5} \to 2,$$

where we interpret line 1 as a loop, multiplying the input by $\frac{10}{3}$ for as long as the product is an integer and when this fails multiplying it by the number $1 = \frac{1}{1}$ and letting control pass to line 2, which multiplies the input by $\frac{3}{5}$ until the result is nonintegral, and then the process stops.

Now the addition sum is performed as before, but after this r_5 is emptied and r_3 filled, leaving the final registers as $r_2 = m + n$,

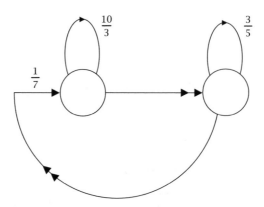

Figure 14.5. Loops which multiply.

$r_3 = m$ and $r_5 = 0$. This is tidier in that it does not destroy the information held in r_3 and uses r_5 as the working register that it really is.

Using a fraction list we can add, but can we multiply? Consider the following program code:

$r_2 := a;$ $r_3 := b;$ $r_7 := c;$
While $r_7 > 0$ do
    Begin
    While $r_3 > 0$ do
        Begin
        $r_2 := r_2 + 1;$ $r_3 := r_3 - 1;$
        End;
    $r_3 := b;$ $r_7 := r_7 - 1;$
    End;
Print r_2; 'is' $a + bc;$

This describes a simple multiplication algorithm for positive integers which we can simulate using figure 14.5. This adds a further loop to figure 14.4, controlled by the fraction $\frac{1}{7}$.

Our pseudo-code is now

$$\text{line 1}: \quad \tfrac{1}{7} \to 2,$$
$$\text{line 2}: \quad \tfrac{10}{3} \to 2, \quad \tfrac{1}{1} \to 3,$$
$$\text{line 3}: \quad \tfrac{3}{5} \to 3, \quad \tfrac{1}{1} \to 1,$$

with the implied interpretation given above.

Notice that the $\frac{1}{1}$ fraction is again used to force a jump.

If we start with $r_2 = a$, $r_3 = b$, $r_5 = 0$, $r_7 = c$, and so with the integer $N = 2^a \times 3^b \times 7^c$, then, after each cycle is completed, we have b added into r_2 and, using r_7 as a counter, we will finish with $r_2 = a + bc$, $r_3 = b$, $r_5 = 0$, $r_7 = 0$. In particular, if we take the case $a = 0$, an input of $N = 3^n \times 7^c$ results in an output of $N = 2^{cn}$, and so multiplication has indeed been accomplished.

Fractran

Where, though, is the fraction list equivalent to this? There is a problem, since the rules force the $\frac{1}{7}$ to be evaluated repeatedly wherever it is encountered in the list, which denies its use as a loop. To manufacture that fraction list we need to look a bit deeper into our pseudo-code, or, as Conway has named it, the Fractran programming language.

Conway defines a Fractran program as a sequence of numbered lines each of the form

$$\text{line } n: \quad \frac{p_1}{q_1} \to n_1, \ \frac{p_2}{q_2} \to n_2, \ \dots, \ \frac{p_r}{q_r} \to n_r,$$

where $n, n_1, n_2, \dots, n_r$ are positive integer line numbers and

$$\frac{p_1}{q_1}, \frac{p_2}{q_2}, \dots, \frac{p_r}{q_r}$$

are fractions. The Fractran machine works by inputting a positive integer N into the lowest numbered line and replacing it by $p_i/q_i \times N$ for the least i for which this is an integer, and then branch to line n_i; if no such integer is possible, the process terminates.

For example, line 10: $\frac{2}{5} \to 15$, $\frac{3}{7} \to 20$ will multiply the input by $\frac{2}{5}$ and change the program flow to line 15 if that product is an integer, otherwise it will multiply the input by $\frac{3}{7}$ and change the program flow to line 20 if that product is an integer, and failing both of these it will stop.

We are interested in fraction lists. In general, the fraction list

$$\left\{ \frac{p_1}{q_1}, \frac{p_2}{q_2}, \dots, \frac{p_r}{q_r} \right\}$$

has the Fractran equivalent of the r lined program:

$$\text{line 1}: \quad \frac{p_1}{q_1} \rightarrow 1, \quad \tfrac{1}{1} \rightarrow 2,$$

$$\text{line 2}: \quad \frac{p_2}{q_2} \rightarrow 2, \quad \tfrac{1}{1} \rightarrow 3,$$

$$\vdots$$

$$\text{line } r: \quad \frac{p_r}{q_r} \rightarrow r,$$

which is an example of what Conway calls a Fractran-r program.
   More compactly, this particular example can be written as the
Fractran-1 program:

$$\text{line 1}: \quad \frac{p_1}{q_1} \rightarrow 1, \frac{p_2}{q_2} \rightarrow 1, \ldots, \frac{p_r}{q_r} \rightarrow 1.$$

   In asking whether our multiplication program can be written
as a fraction list, we are asking whether this Fractran-3 program
can be written as a Fractran-1 program. In fact, in his article Conway demonstrates a method which allows an arbitrary Fractran-r
program to be simulated by a Fractran-1 program, and therefore
a fraction list; it uses the unique factorization property of the
primes:

(1) Clear the program of all loops.

(2) Label the nodes of the diagram with distinct primes $P, Q,$
    $R, \ldots$ larger than any primes appearing in the numerators
    or denominators of its fractions: these will form the new
    line numbers.

(3) Translate, in order, a line at a time by using the scheme

$$\text{line } P: \quad \frac{p_1}{q_1} \rightarrow Q, \frac{p_2}{q_2} \rightarrow R, \ldots \quad \text{goes to} \quad \left\{ \frac{p_1 Q}{q_1 P}, \frac{p_2 R}{q_2 P}, \ldots \right\}.$$

(4) Populate the fraction list in the correct order.

In the case of the multiplication he redesigned figure 14.5 to
figure 14.6.

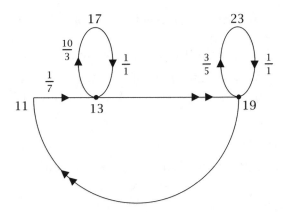

Figure 14.6. The multiplication loops modified.

Now rewrite the line numbers accordingly to get the Fractran program:

line 11 : $\frac{1}{7} \rightarrow 13$,

line 13 : $\frac{10}{3} \rightarrow 17$, $\frac{1}{1} \rightarrow 19$,

line 17 : $\frac{1}{1} \rightarrow 13$,

line 19 : $\frac{3}{5} \rightarrow 23$, $\frac{1}{1} \rightarrow 11$,

line 23 : $\frac{1}{1} \rightarrow 19$,

and use the algorithm to generate the fraction list

$$\{\tfrac{13}{77}, \tfrac{170}{39}, \tfrac{13}{17}, \tfrac{19}{13}, \tfrac{69}{95}, \tfrac{19}{23}, \tfrac{11}{19}\}.$$

Notice the order of the fractions, with the second entries on the two lines fitting into their proper places.

The reader can check that starting with $r_2 = a$, $r_3 = b$, $r_5 = 0$, $r_7 = c$, $r_{11} = 1$ and hence $N = 2^a \times 3^b \times 7^c \times 11$ results in $r_2 = a + bc$, $r_3 = b$, $r_5 = 0$, $r_7 = 0$, $r_{11} = 1$ and hence $N = 2^{a+bc} \times 3^b \times 11$.

We can interpret these prime node labels as states of the Fractran machine. Doing so and disassembling the fraction as

$$\left\{ \begin{array}{ccccccc} \frac{13}{7\times 11} & \frac{2\times 5\times 17}{3\times 13} & \frac{13}{17} & \frac{19}{13} & \frac{3\times 23}{5\times 19} & \frac{19}{23} & \frac{11}{19} \\ A & B & C & D & E & F & G \end{array} \right\}$$

enables us to interpret each fraction in the following way:

A: $11 \rightarrow 13$ with a multiplier of $\frac{1}{7}$, which causes $r_7 \rightarrow r_7 - 1$;

B: $13 \rightarrow 17$ with a multiplier of $\frac{2 \times 5}{3}$, which causes
$\qquad r_2 \rightarrow r_2 + 1, \ r_5 \rightarrow r_5 + 1, \ r_7 \rightarrow r_7 - 1$;

C: $17 \rightarrow 13$;

D: $13 \rightarrow 19$;

E: $19 \rightarrow 23$ with a multiplier of $\frac{3}{5}$, which causes
$\qquad r_3 \rightarrow r_3 + 1, \ r_5 \rightarrow r_5 - 1$;

F: $23 \rightarrow 19$;

G: $19 \rightarrow 11$;

and we have a machine which moves between states, possibly altering the contents of certain dynamic registers.

Fibonnaci Game

As a final precursor to dealing with PRIMEGAME, we will look at FIBONNACIGAME and so go through the procedure that generates any particular Fibonnaci number.

The Fibonnaci sequence is well known to be defined by the recurrence relation:

$$a_1 = a_2 = 1 \quad \text{and} \quad a_{n+2} = a_n + a_{n+1} \quad \text{for } n \geqslant 1,$$

which generates $1, 1, 2, 3, 5, 8, 13, \ldots$. To have a computer print out the nth Fibonnaci number requires code equivalent to

```
r2 := 1;   r3 := 1;   r5 := n;   r13 := 1;
While r5 > 0 do
    Begin
        r7 := r2 + r3;   r2 := r3;   r3 := r7;   r5 := r5 - 1;
    End;
Print r7;
```

Figure 14.7 represents the process and, with an input of $N = 2 \times 3 \times 5^n \times 13$, the Fractran-5 program to achieve this is given

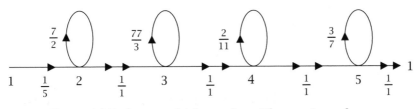

Figure 14.7. Loops which produce Fibonacci numbers.

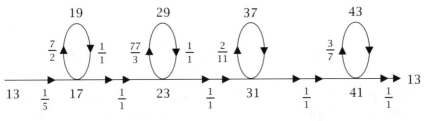

Figure 14.8. Modified Fibonacci loops.

below:

line 1 : $\frac{1}{5} \to 2 \ldots$ this subtracts 1 from the 5 register, which is the counter;

line 2 : $\frac{7}{2} \to 2, \frac{1}{1} \to 3 \ldots$ this copies Fib(r) into register 7;

line 3 : $\frac{77}{3} \to 3, \frac{1}{1} \to 4 \ldots$ this adds Fib$(r+1)$ into register 7 and also copies it into register 11;

line 4 : $\frac{2}{11} \to 4, \frac{1}{1} \to 5 \ldots$ Fib(r) is replaced by Fib$(r+1)$;

line 5 : $\frac{3}{7} \to 5, \frac{1}{1} \to 1 \ldots$ continue the process.

To convert this to a fraction list we need to eliminate the loops and label the nodes with 'big' primes, as in figure 14.8, which enables us to write the Fractran-9 program as

line 13 : $\frac{1}{5} \to 17,$

line 17 : $\frac{7}{2} \to 19, \frac{1}{1} \to 23,$

line 19 : $\frac{1}{1} \to 17,$

line 23 : $\frac{77}{3} \to 29, \frac{1}{1} \to 31,$

line 29 : $\frac{1}{1} \to 23,$

line 31 : $\frac{2}{11} \to 37, \frac{1}{1} \to 41,$

$$\text{line } 37: \quad \tfrac{1}{1} \to 31,$$

$$\text{line } 41: \quad \tfrac{3}{7} \to 43, \ \tfrac{1}{1} \to 13,$$

$$\text{line } 43: \quad \tfrac{1}{1} \to 41,$$

and using Conway's algorithm to finish with the fraction list

$$\left\{\tfrac{17}{65}, \tfrac{133}{34}, \tfrac{17}{19}, \tfrac{23}{17}, \tfrac{2233}{69}, \tfrac{23}{29}, \tfrac{31}{23}, \tfrac{74}{341}, \tfrac{31}{37}, \tfrac{41}{31}, \tfrac{129}{287}, \tfrac{41}{43}, \tfrac{13}{41}, \tfrac{1}{13}, \tfrac{1}{3}\right\}$$

with the final two fractions there to tidy things up so that the process results in the output of precisely $2^{\text{Fib}(n)}$. These final fractions are the equivalent of including one further node, labelled with a 1 and with two loops attached to it. Although the use of a nonprime labelled node and a loop does challenge the constraints of the process, it is fine used in this way, and, in Conway's words:

> We note that it is permissible to label one of the states with the number 1, rather than a large prime number. The fractions corresponding to transitions from this state should be placed (in their proper order) at the end of the Fractran-1 program. If this is done, loops, provided they have lower priority than any other transition, are permitted at node 1.

He demonstrates the point by amending the fraction list for multiplication to

$$\left\{\tfrac{170}{39}, \tfrac{19}{13}, \tfrac{13}{17}, \tfrac{69}{95}, \tfrac{1}{19}, \tfrac{13}{7}, \tfrac{1}{3}\right\},$$

which we leave the reader to ponder!

Finally, to PRIMEGAME.

PRIMEGAME

In factored form its fraction list is

$$\left\{ \begin{array}{ccccccc}
\frac{17}{7\times13} & \frac{2\times3\times13}{5\times17} & \frac{19}{3\times17} & \frac{23}{2\times19} & \frac{29}{3\times11} & \frac{7\times11}{29} & \frac{5\times19}{23} \\
A & B & C & D & E & F & G
\end{array} \right.$$

$$\left. \begin{array}{ccccccc}
\frac{7\times11}{19} & \frac{1}{17} & \frac{11}{13} & \frac{13}{11} & \frac{3\times5}{2\times7} & \frac{3\times5}{2} & \frac{5\times11}{1} \\
H & I & J & K & L & M & N
\end{array} \right\},$$

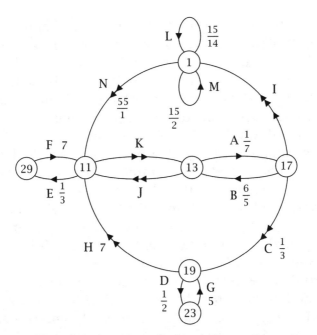

Figure 14.9. Loops which produce primes.

which can be written as the Fractran-7 program:

line 13 : $\frac{1}{7} \to 17, \frac{1}{1} \to 11$;

line 17 : $\frac{6}{5} \to 13, \frac{1}{3} \to 19, \frac{1}{1} \to 1$;

line 19 : $\frac{1}{2} \to 23, 7 \to 11$;

line 11 : $\frac{1}{3} \to 29, \frac{1}{1} \to 13$;

line 29 : $7 \to 11$;

line 23 : $5 \to 19$;

line 1 : $\frac{15}{14} \to 1, \frac{15}{2} \to 1, 5 \to 11$.

Figure 14.9 represents the process and is altogether more complicated. Its nodes are labelled with the obvious primes and with 1.

At the sixth stage, PRIMEGAME takes $N = 2$ to $N = 2275$, the factorization of which is $5^2 \times 7 \times 13$, at which point N is subject to the (AB) cycle. This is a special case of the general $N = 5^n \times 7^d \times 13$ with $d < n$, which will recur throughout the process and figure 14.8 shows what inevitably happens to such a number. It is transformed to a number of the form $N = 2^n \times 3^r \times 7^{d-r-1} \times 17$ and then, if $r > 0$, it proceeds via the C route to $5^n \times 7^{d-1} \times 13$, whereas, if $r = 0$, it proceeds via the I route to $5^{n+1} \times 7^n \times 13$. Now we can interpret this in terms of the integer n and its possible divisors d by writing $n = q \times d + r$ in the standard way. PRIMEGAME acting on $N = 5^n \times 7^{n-1} \times 13$, as it does initially with $n = 2$, will test all possible divisors of n from $n - 1$ to 1, and then continues with n increased by 1. But, if $r = 0$, it is the case that d divides n and therefore n will be composite unless $d = 1$, in which case n will be prime. The number $2^n \times 7^{d-1} = 2^n$ then provides the only power of 2 that ever arises in the computation, and does so precisely when n is prime. Even unveiled, it's still clever!

Richard Guy has established the fact in his own way by producing the flowchart for the process. Writing the contents of the 2 and 5 registers as t and r, respectively, and realizing that $t + r = n$ and also writing the contents of the 3 and 7 registers as s and q and realizing that $s + q = d$, he produced figure 14.10, which is equivalent to figure 14.9.

In 1999 Devin Kilminster of the University of Western Australia gave a talk on how Conway's fourteen fractions can be reduced to the ten; those fractions are

$$\left\{ \frac{7}{3}, \frac{99}{98}, \frac{13}{49}, \frac{39}{35}, \frac{36}{91}, \frac{10}{143}, \frac{49}{13}, \frac{7}{11}, \frac{1}{2}, \frac{91}{1} \right\},$$

where the initial value for N is 10 and the primes are generated by subsequent powers of 10.

Of course, this is a theoretical process. We have seen with PRIMEGAME just how many steps it takes to generate the first few primes and, in fact, Richard Guy gave the following formula for the number of steps needed to inspect the number n for primeness:

$$n - 1 + (6n + 2)(n - b) + 2 \sum_{d=b}^{n-1} \left\lfloor \frac{n}{d} \right\rfloor,$$

where $b < n$ is the biggest divisor of n; for prime n, b is, of course, 1.

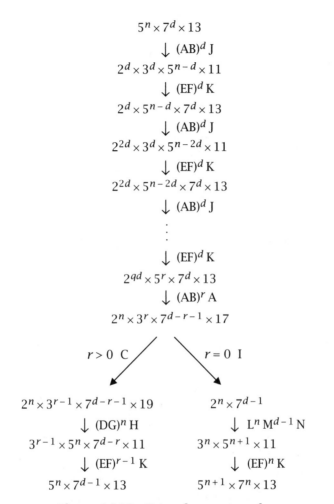

$$5^n \times 7^d \times 13$$
$$\downarrow (AB)^d J$$
$$2^d \times 3^d \times 5^{n-d} \times 11$$
$$\downarrow (EF)^d K$$
$$2^d \times 5^{n-d} \times 7^d \times 13$$
$$\downarrow (AB)^d J$$
$$2^{2d} \times 3^d \times 5^{n-2d} \times 11$$
$$\downarrow (EF)^d K$$
$$2^{2d} \times 5^{n-2d} \times 7^d \times 13$$
$$\downarrow (AB)^d J$$
$$\vdots$$
$$\downarrow (EF)^d K$$
$$2^{qd} \times 5^r \times 7^d \times 13$$
$$\downarrow (AB)^r A$$
$$2^n \times 3^r \times 7^{d-r-1} \times 17$$

$r > 0$ C $\qquad\qquad$ $r = 0$ I

$$2^n \times 3^{r-1} \times 7^{d-r-1} \times 19 \qquad\qquad 2^n \times 7^{d-1}$$
$$\downarrow (DG)^n H \qquad\qquad\qquad \downarrow L^n M^{d-1} N$$
$$3^{r-1} \times 5^n \times 7^{d-r} \times 11 \qquad\qquad 3^n \times 5^{n+1} \times 11$$
$$\downarrow (EF)^{r-1} K \qquad\qquad\qquad \downarrow (EF)^n K$$
$$5^n \times 7^{d-1} \times 13 \qquad\qquad\qquad 5^{n+1} \times 7^n \times 13$$

Figure 14.10. Prime loops at work.

To form some idea of just how this accumulates, we can follow Guy's thoughts as he answered Conway's own question about how many steps are needed for PRIMEGAME to generate the thousandth prime (8831). To answer this we must compute the sum of the expression from $n = 2$ to $n = 8831$, which results in the number $1\,378\,197\,377\,195 \approx 1.4 \times 10^{12}$.

Perhaps this is slow, but a look at PIGAME puts this into perspective. Starting with $N = 89 \times 2^n$, the following fraction list computes the nth digit of $\pi = 3.141\,59\ldots$; that is, it stops with

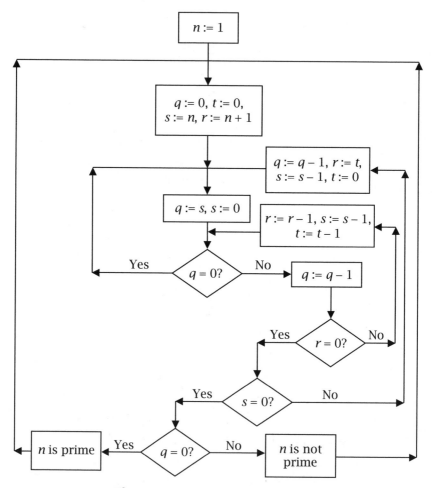

Figure 14.11. The Guy flowchart.

$2^{\pi(n)}$:

$$\left\{ \frac{365}{46}, \frac{29}{161}, \frac{79}{575}, \frac{679}{451}, \frac{3159}{413}, \frac{83}{407}, \frac{473}{371}, \frac{638}{355}, \frac{434}{335}, \frac{89}{235}, \frac{17}{209}, \frac{79}{122}, \right.$$

$$\frac{31}{183}, \frac{41}{115}, \frac{517}{89}, \frac{111}{83}, \frac{305}{79}, \frac{23}{73}, \frac{73}{71}, \frac{61}{67}, \frac{37}{61}, \frac{19}{59}, \frac{89}{57}, \frac{41}{53}, \frac{833}{47}, \frac{53}{43},$$

$$\left. \frac{86}{41}, \frac{13}{38}, \frac{23}{37}, \frac{67}{31}, \frac{71}{29}, \frac{83}{19}, \frac{475}{17}, \frac{59}{13}, \frac{41}{291}, \frac{1}{7}, \frac{1}{11}, \frac{1}{1024}, \frac{1}{97}, \frac{89}{1} \right\}.$$

More explicitly, with $n = 0$ the machine stops at 3, with $n = 1$ the machine stops at 1, $n = 2$ the machine stops at 4, and so on. This is all very well, but Bill Dubuque has commented that

PIGAME computes the nth digit of π by using over $4 \times 2^{10^n}$ terms of Wallis's product ($\frac{1}{2}\pi = \frac{2}{1} \times \frac{2}{3} \times \frac{4}{3} \times \frac{4}{5} \times \frac{6}{5} \times \frac{6}{7} \times \cdots$), which makes it in practice unrealistic for $n > 1$.

In the end, PRIMEGAME is simply a striking example of the programming language Fractran, which Conway has shown capable of simulating any computable process, just as his famous game of Life is capable of doing.

The Motifs

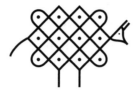

Fast cars, fast women, fast algorithms... what more could a man want?

Joe Mattis

Ethnomathematics

Southwest Africa finds its most celebrated place on the mathematical map on the border of Uganda and Zaire, since it was there in 1960 that the Belgian geologist Jean de Heinzelin discovered on the shores of Lake Edward the ancient Ishango Bone; its provenance is disputed with its date varying from 8000 to 20 000 B.C.E. and its purpose from a lunar calendar to a list of prime numbers. Yet there are other African ethnomathematical treasures, and the attractive designs which have featured at the start of each of the book's chapters point to one such: sona, or in singular lusona. These are examples of a small but rich part of the cultural heritage of the Chokwe (pronounced *Chockway*) group of the Bantu people of northeast Angola (whose lands spill into Zambia and Zaire). The Chokwe are renowned for their figurative and decorative art with sona integrating this art with their wider culture, and also with mathematics. We will take a brief look at that quite surprising mathematical connection.

The Mathematics of the Motifs

To construct a design, imagine the dots to be one unit apart and a rectangular grid of them surrounded by a rectangle of mirrors extending half a unit outside them. The construction algorithm is:

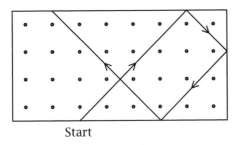

Start

Figure 1. Dots, mirrors and a path.

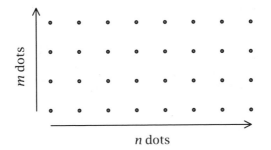

n dots

Figure 2. The general array of dots.

- Start a line on a mirror, directly below (above or to the side of) a dot and continue it at 45° until the opposite mirror line is reached.
- Reflect the line through 90° and continue extending the line in this way.
- If the line returns to its beginning before every dot has been enclosed, start a second line near another unenclosed peripheral dot.
- Repeat until all dots have been enclosed.
- Smooth the final figure if desired.

The layout and start of the procedure is shown in figure 1.

To analyse the process we adopt the following notation: $f(m, n)$ is the number of separate paths needed to enclose all dots of the rectangular array shown in figure 2.

It is evident that $f(1, n) = 1$, and we have seen examples of this in the motifs which are at the start of the Introduction and chapters 1 and 2. That $f(m, n) = f(n, m)$ is also evident.

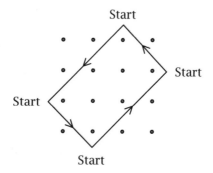

Figure 3. A complete path in a square array of dots.

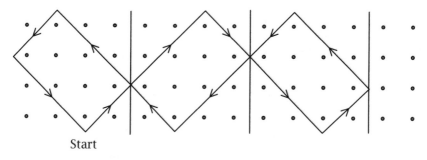

Start

Figure 4. A path in a rectangular array of dots.

To enable us to recognize the general form of $f(m, n)$ we first examine the special case of a square array of size $n \times n$. By symmetry the path that starts near a particular dot at a side of the array returns to that side at the same place, as indicated in figure 3; this means that, if we have to enclose all dots on a side (and therefore all dots), we need a path for each one of them and this means that $f(n, n) = n$.

Now we look at the general case of an $m \times n$ array of dots and suppose that $m < n$. Figure 4 shows such an array, where the vertical lines divide it into $m \times m$ squares and what rectangle (if any) is left over. Wherever we start the design in the leftmost square, allowing the path to enter the remaining square arrays adds to the design but in no way alters the reflected direction of the path; again, this is shown in figure 4. This means that, to count the paths needed to enclose all of the dots, we can remove these square arrays from consideration; notationally, if

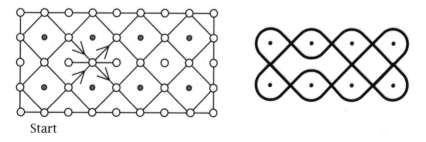

Start

Figure 5. The 2×4 array with an added two-way mirror.

we write $n = qm + r$ (with $r < m$), then $f(m, n) = f(m, m + r)$ and since, by the same reasoning, the first square makes no difference, $f(m, n) = f(m, r)$.

Now let us take two (rather large) numeric examples which utilize these several observations:

$$f(144, 2068) = f(144, 52) = f(52, 144) = f(52, 40)$$
$$= f(40, 52) = f(40, 12) = f(12, 40)$$
$$= f(12, 4) = f(4, 12) = f(4, 4) = 4$$

and

$$f(123, 2113) = f(123, 22) = f(22, 123) = f(22, 13)$$
$$= f(13, 22) = f(13, 9) = f(9, 13)$$
$$= f(9, 4) = f(4, 9) = f(4, 1) = f(1, 4) = 1.$$

By now we hope that the reader will have recognized the workings of the Euclidean Algorithm, which is used to find the highest common factor of two integers: in short,

$$f(m, n) = \text{HCF}(m, n).$$

The motifs which appear at the start of the chapters are drawn with lines of various hues where necessary to indicate the number needed in each case to complete the design.

The Reality of the Construction

This may be interesting mathematically but it is of no use to the Chokwe: sona have their primary place in the culture as

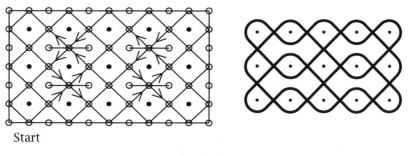

Start

Figure 6.

figurative representations (as in the motif at the top of this section and in the front matter of the book; the former is styled
'an antelope' and the latter 'an antelope's footprint') and also
mnemonics for stories or lessons which are important to their
folk law. The diagrams are usually created dynamically by the
male Chokwe (to be exact, the *akwa kuta sona* or *those who know
how to draw*), who trace lines with their fingers in the flat sand,
after having formed the dots with the finger tips. As the story
unfolds, so the motif is continuously built up as a single curve,
the story finishing as the design is completed: it is not permitted
to stop and then start another curve, indeed, pausing is frowned
upon; the skills are passed from generation to generation during
a six- to eight-month period of male initiation rites. If the HCF of
the dimensions of the rectangle is not 1, we have seen that the
diagram cannot be traced by one continuous movement and this
contradicts this essential requirement of the process. Of course,
such rectangles could be abandoned but an alternative strategy is to introduce small two-way mirror lines within the array,
placed to prevent this happening: this has the added advantage
of enhancing the designs further and can be applied whether or
not the dimensions are coprime. Look, for example, at figure 5,
the left part of which shows the 2×4 shape which is the basis of
the motif from chapter 5, to which has been added a small horizontal two-sided mirror. If the path is traced out from the start
position, we can see that the mirror line prevents any premature
closing up: all of the dots are encircled by one continuous path.
The resulting, smoothed pattern is shown on the right.

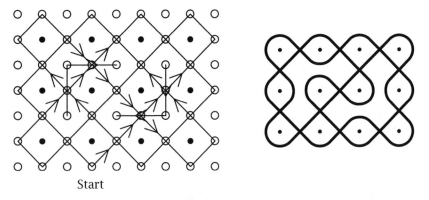

Figure 7.

As another example of the effect consider the 3×5 array which appeared at the start of chapter 10 and which, of course, can be traced in one continuous loop. Its dot pattern, to which has been added four small horizontal, two-way mirrors, is shown on the left of figure 6. The resulting smoothed motif is, again, on the right; it and its bigger versions are given the name 'Lion's Stomach'.

The 3×4 array of chapter 9 can be enhanced by the inclusion of two sets of two-way mirrors set at right angles to each other, as shown on the left of figure 7; again, the smoothed version is shown on the right of the figure. This is the 'Chased Chicken' design.

These motifs, then, provide a link between art, African culture and mathematics, with the Euclidean Algorithm making a somewhat surprising appearance. For further exploration of these and other related ideas the reader is referred to the works of Paul Gerdes, starting with *Geometry from Africa: Mathematical and Educational Explorations*, published by The Mathematical Association of America.

Appendix A

THE INCLUSION–EXCLUSION PRINCIPLE

This is used to count the distinct elements in any number of overlapping sets. To begin with, consider the Venn diagram in figure A.1(a), in which just two sets intersect.

To calculate the total number of elements we can add all of those elements in set A to all of those in set B, but in doing so we have counted those in the intersection twice and so we subtract one count of the elements in the intersection. Put symbolically,

$$n(A \cup B) = n(A) + n(B) - n(A \cap B).$$

With the three overlaps, as in figure A.1(b), we adopt the same process, this time subtracting all elements in the three intersections, but in doing this we have now eliminated those in the intersection of all three sets. If we add these back in, we arrive at the expression

$$n(A \cup B \cup C) = n(A) + n(B) + n(C) - n(A \cap B)$$
$$- n(A \cap C) - n(B \cap C) + n(A \cap B \cap C).$$

In general,

$$n\left(\bigcup_{i=1}^{n} A_i\right) = \sum_{i=1}^{n} n(A_i) - \sum_{i,j:i<j}^{n} n(A_i \cap A_j)$$
$$+ \sum_{i,j,k:i<j<k}^{n} n(A_i \cap A_j \cap A_k) - \cdots$$
$$\pm n(A_1 \cap A_2 \cap A_3 \cdots \cap A_n).$$

We can convince ourselves that the process does count every element precisely once by the following argument.

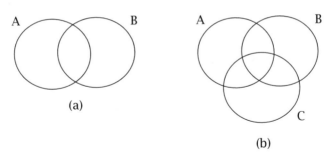

(a)

(b)

Figure A.1.

Suppose that an element x belongs to exactly r of the sets. The first step has us count all of the elements in all of the sets, disregarding overlap; this means that x is counted exactly r times. There are $\binom{r}{2}$ pairs of the sets in which x appears, so we need to subtract this number from r; then we need to add in the number of times that x appears in set triples, which is $\binom{r}{3}$, and continue doing this until we reach $\binom{r}{r}$, which is 1. Therefore, the process counts x precisely

$$r - \binom{r}{2} + \binom{r}{3} - \binom{r}{4} + \cdots \binom{r}{r}$$

times.

Now notice that the expression

$$-1 + r - \binom{r}{2} + \binom{r}{3} - \binom{r}{4} + \cdots \binom{r}{r}$$

$$= -\left[1 - r + \binom{r}{2} - \binom{r}{3} + \binom{r}{4} - \cdots \binom{r}{r} \right]$$

$$= -(1 + (-1))^r = 0$$

and so

$$r - \binom{r}{2} + \binom{r}{3} - \binom{r}{4} + \cdots \binom{r}{r} = 1,$$

and the process has indeed counted x precisely once.

Appendix B

THE BINOMIAL INVERSION FORMULA

Suppose that we have a function $f(r, s)$ of two positive integer variables and that we wish to sum its values over the infinite, diagonal half-plane:

$$f(0,0) \qquad f(1,0) \qquad f(2,0) \qquad f(3,0) \qquad \cdots$$
$$f(1,1) \qquad f(2,1) \qquad f(3,1) \qquad \cdots$$
$$f(2,2) \qquad f(3,2) \qquad \cdots$$
$$f(3,3) \qquad \cdots$$
$$\vdots \qquad \vdots$$

We could add all of the terms together in two obvious ways:

(1) Take each row one at a time and add the terms in it and then add the sums of these rows. The sum of the sth row is

$$f(s, s) + f(s + 1, s) + f(s + 2, s) + \cdots$$

and adding these contributions from each row is achieved by

$$\sum_{s=0}^{\infty} \{f(s, s) + f(s + 1, s) + f(s + 2, s) + \cdots\}$$

and this sum can be compactly written as the double sum $\sum_{s=0}^{\infty} \sum_{r=s}^{\infty} f(r, s)$.

(2) Take each column one at a time and add the terms in it and then add the sums of these columns. The sum of the rth column is

$$f(r, 0) + f(r, 1) + f(r, 2) + \cdots + f(r, r),$$

where each sum is finite, ending at the appropriate diagonal element. Adding these contributions from each column is achieved by

$$\sum_{r=0}^{\infty} \{f(r,0) + f(r,1) + f(r,2) + \cdots + f(r,r)\}$$

and this sum can be compactly written as the double sum $\sum_{r=0}^{\infty} \sum_{s=0}^{r} f(r,s)$

Put these observations together and we have the identity

$$\sum_{r=0}^{\infty} \sum_{s=0}^{r} f(r,s) = \sum_{s=0}^{\infty} \sum_{r=s}^{\infty} f(r,s).$$

Now for the formula.

If two sets of numbers

$$\{a_0, a_1, a_2, \ldots, a_n\} \quad \text{and} \quad \{b_0, b_1, b_2, \ldots, b_n\}$$

are related by the condition

$$b_n = \sum_{r=0}^{n} \binom{n}{r} a_r,$$

then

$$a_n = \sum_{r=0}^{n} (-1)^{n-r} \binom{n}{r} b_r.$$

The plan is to define two generating functions $A(x)$ and $B(x)$ by

$$A(x) = \sum_{r=0}^{\infty} \frac{a_r x^r}{r!} \quad \text{and} \quad B(x) = \sum_{r=0}^{\infty} \frac{b_r x^r}{r!}$$

and write $B(x)$ in terms of $A(x)$. This will allow us to accomplish the reverse identity of writing $A(x)$ in terms of $B(x)$ and this in turn will enable $\{a_0, a_1, a_2, \ldots, a_n\}$ to be written in terms of $\{b_0, b_1, b_2, \ldots, b_n\}$.

Using the definition of the b_r and substituting them into the definition of $B(x)$ results in

$$B(x) = \sum_{r=0}^{\infty} \left(\sum_{s=0}^{r} \binom{r}{s} a_s \right) \frac{x^r}{r!}$$

$$= \sum_{r=0}^{\infty} \sum_{s=0}^{r} \frac{r!}{(r-s)!\,s!} a_s \frac{x^r}{r!}$$

$$= \sum_{r=0}^{\infty} \sum_{s=0}^{r} \frac{a_s}{(r-s)!\,s!} x^r$$

$$= \sum_{r=0}^{\infty} \sum_{s=0}^{r} \left(\frac{a_s x^s}{s!} \right) \left(\frac{x^{r-s}}{(r-s)!} \right).$$

Now we invoke the earlier result with

$$f(r,s) = \left(\frac{a_s x^s}{s!} \right) \left(\frac{x^{r-s}}{(r-s)!} \right)$$

to arrive at

$$B(x) = \sum_{s=0}^{\infty} \sum_{r=s}^{\infty} \left(\frac{a_s x^s}{s!} \right) \left(\frac{x^{r-s}}{(r-s)!} \right).$$

Now we can move the summation over r past the expression in s to get

$$B(x) = \sum_{s=0}^{\infty} \left(\frac{a_s x^s}{s!} \right) \sum_{r=s}^{\infty} \left(\frac{x^{r-s}}{(r-s)!} \right).$$

Next we clean up the second summation by writing $t = r - s$ and so replacing the variable r by the variable t to get

$$B(x) = \sum_{s=0}^{\infty} \left(\frac{a_s x^s}{s!} \right) \sum_{t=0}^{\infty} \left(\frac{x^t}{t!} \right).$$

The first summation is $A(x)$ and the second simply e^x, which makes

$$B(x) = A(x)e^x.$$

Now we can reverse the identity to get $A(x) = e^{-x}B(x)$ and, in turn, this means that

$$A(x) = e^{-x}B(x) = \left(\sum_{r=0}^{\infty} \frac{(-1)^r x^r}{r!} \right) \left(\sum_{s=0}^{\infty} \frac{b_s x^s}{s!} \right)$$

$$= \sum_{r=0}^{\infty} \sum_{s=0}^{\infty} \frac{(-1)^r x^r}{r!} \frac{b_s x^s}{s!}$$

$$= \sum_{s=0}^{\infty} \sum_{r=0}^{\infty} \frac{(-1)^r x^r}{r!} \frac{b_s x^s}{s!}.$$

To alter this to a convenient form, replace r by $n = r + s$ to get

$$A(x) = \sum_{s=0}^{\infty} \sum_{n=s}^{\infty} \frac{(-1)^{n-s} x^{n-s}}{(n-s)!} \frac{b_s x^s}{s!}$$

and using the preliminary result once more and introducing $n!$ at the top and bottom we have

$$A(x) = \sum_{n=0}^{\infty} \sum_{s=0}^{n} \frac{(-1)^{n-s} x^{n-s}}{(n-s)!} \frac{b_s x^s}{s!}$$

$$= \sum_{n=0}^{\infty} \sum_{s=0}^{n} \frac{n!}{(n-s)! \, s!} (-1)^{n-s} \frac{x^n}{n!} b_s.$$

Tidying up and pushing the sigma through results in

$$A(x) = \sum_{n=0}^{\infty} \frac{x^n}{n!} \sum_{s=0}^{n} \binom{n}{s} (-1)^{n-s} b_s.$$

Therefore,

$$A(x) = \sum_{n=0}^{\infty} \frac{x^n}{n!} a_n = \sum_{n=0}^{\infty} \frac{x^n}{n!} \left[\sum_{s=0}^{n} \binom{n}{s} (-1)^{n-s} b_s \right].$$

And, equating coefficients, we finally have the result that

$$a_n = \sum_{s=0}^{n} \binom{n}{s} (-1)^{n-s} b_s.$$

Appendix C

SURFACE AREA AND ARC LENGTH

The element of surface area generated by rotating the elemental piece of curve 360° around the x-axis, as shown in figure C.1, is given by

$$\delta S \approx 2\pi y \times \sqrt{(\delta x)^2 + (\delta y)^2}$$

and so

$$\frac{\delta S}{\delta x} \approx \frac{2\pi y \times \sqrt{(\delta x)^2 + (\delta y)^2}}{\delta x} = 2\pi y \sqrt{1 + \left(\frac{\delta y}{\delta x}\right)^2}$$

and in the limit

$$\frac{dS}{dx} = 2\pi y \sqrt{1 + \left(\frac{dy}{dx}\right)^2},$$

which makes the total surface area

$$S = 2\pi \int y \sqrt{1 + \left(\frac{dy}{dx}\right)^2}\, dx.$$

Arc Length of a Curve

Take an arbitrary smooth curve and an origin O, then in the standard calculus notation of figure C.2, and using Pythagoras's Theorem in the upper pseudo-triangular, elemental region we get

$$(\delta s)^2 \approx (\delta r)^2 + (r\delta\theta)^2.$$

We can choose to divide both sides by δr^2 to get

$$\left(\frac{\delta s}{\delta r}\right)^2 \approx 1 + \left(r\frac{\delta\theta}{\delta r}\right)^2$$

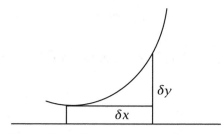

Figure C.1. Element for surface area.

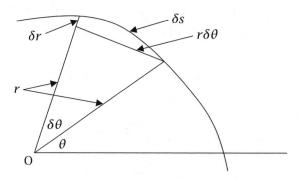

Figure C.2. Element for arc length.

and in the limit

$$\left(\frac{\mathrm{d}s}{\mathrm{d}r}\right)^2 = 1 + \left(r\frac{\mathrm{d}\theta}{\mathrm{d}r}\right)^2,$$

which makes

$$s = \int \sqrt{1 + \left(r\frac{\mathrm{d}\theta}{\mathrm{d}r}\right)^2}\,\mathrm{d}r.$$

Alternatively, we can choose to divide both sides by $\delta\theta^2$ to get

$$\left(\frac{\delta s}{\delta\theta}\right)^2 \approx \left(\frac{\delta r}{\delta\theta}\right)^2 + r^2$$

and in the limit

$$\left(\frac{\mathrm{d}s}{\mathrm{d}\theta}\right)^2 = \left(\frac{\mathrm{d}r}{\mathrm{d}\theta}\right)^2 + r^2,$$

which makes

$$s = \int \sqrt{\left(\frac{\mathrm{d}\theta}{\mathrm{d}r}\right)^2 + r^2}\,\mathrm{d}\theta.$$

Index